~~smarter~~ *faster* **better**

Charles Duhigg is a Pulitzer Prize-winning investigative reporter for *The New York Times* and the author of *The Power of Habit*. He is a winner of the National Academies of Sciences, National Journalism, and George Polk awards. A graduate of Harvard Business School and Yale College, he lives in Brooklyn with his wife and two children.

ALSO BY CHARLES DUHIGG

The Power of Habit

Charles Duhigg

smarter
faster
better

WILLIAM HEINEMANN: LONDON

1 3 5 7 9 10 8 6 4 2

William Heinemann
20 Vauxhall Bridge Road
London SW1V 2SA

William Heinemann is part of the Penguin Random House
group of companies whose addresses can be found
at global.penguinrandomhouse.com.

Penguin
Random House
UK

First published in the United States by Random House Books in 2016.

www.randomhouse.co.uk

A CIP catalogue record for this book is available from the British Library.

ISBN 9780434023455 (Hardback)
ISBN 9780434023462 (Trade paperback)

Printed and bound by Clays Ltd, St Ives plc

Penguin Random House is committed to a sustainable
future for our business, our readers and our planet. This book is
made from Forest Stewardship Council® certified paper.

MIX
Paper from
responsible sources
FSC® C018179

To Harry, Oliver,
Doris and John,
Andy,
and, most of all, Liz

CONTENTS

smarter
faster
better

INTRODUCTION

My introduction to the science of productivity began in the summer of 2011, when I asked a friend of a friend for a favor.

At the time, I was finishing a book about the neurology and psychology of habit formation. I was in the final, frantic stages of the writing process—a flurry of phone calls, panicked rewrites, last-minute edits—and felt like I was falling farther and farther behind. My wife, who worked full-time, had just given birth to our second child. I was an investigative reporter at *The New York Times* and spent my days chasing stories and my nights rewriting book pages. My life felt like a treadmill of to-do lists, emails requiring immediate replies, rushed meetings, and subsequent apologies for being late.

Amid all this hustle and scurry—and under the guise of asking for a little publishing advice—I sent a note to an author I admired, a friend of one of my colleagues at the *Times*. The author's name was Atul Gawande, and he appeared to be a paragon of success. He was a forty-six-year-old staff writer at a prestigious magazine, as well as a renowned surgeon at one of the nation's top hospitals. He was an associate professor at Harvard, an adviser to the World Health Organization, and the founder of a nonprofit that sent surgical supplies to medically underserved parts of the world. He had written three books—all bestsellers—and was married with three children. In 2006, he had been awarded a MacArthur "genius" grant—and had promptly given a substantial portion of the $500,000 prize to charity.

There are some people who pretend at productivity, whose résumés appear impressive until you realize their greatest talent is self marketing. Then there are others, like Gawande, who seem to exist on a different plane of getting things done. His articles were smart and engaging, and, by all accounts, he was gifted in the operating room, committed to his patients, and a devoted father. Whenever he was interviewed on television, he appeared relaxed and thoughtful. His accomplishments in medicine, writing, and public health were important and real.

I emailed him to ask if he had some time to talk. I wanted to know how he managed to be so productive. Mainly, what was his secret? And, if I learned it, could I change my own life?

"Productivity," of course, means different things in different settings. One person might spend an hour exercising in the morning before dropping the kids at school and consider the day a success. Another might opt to use that time locked in her office, returning emails and calling a few clients, and feel equally accomplished. A research scientist or artist may see productivity in failed experiments or discarded canvases since each mistake, they hope, gets them closer to discovery, while an engineer's measure of productivity might focus on making an assembly line ever faster. A productive weekend might involve walking through the park with your kids, while a productive workday involves rushing them to daycare and getting to the office as early as you can.

Productivity, put simply, is the name we give our attempts to figure out the best uses of our energy, intellect, and time as we try to seize the most meaningful rewards with the least wasted effort. It's a process of learning how to succeed with less stress and struggle. It's about getting things done without sacrificing everything we care about along the way.

By this definition, Atul Gawande seemed to have things pretty well figured out.

A few days later, he responded to my email with his regrets. "I wish I could help," he wrote, "but I'm running flat out with my various commitments." Even he, it seemed, had limits. "I hope you'll understand."

Later that week, I mentioned this exchange to our mutual friend. I made it clear I wasn't offended—that, in fact, I admired Gawande's focus. I imagined his days were consumed with healing patients, teaching medical students, writing articles, and advising the world's largest health organization.

No, my friend told me, I had it wrong. That wasn't it. Gawande was particularly busy that week because he had bought tickets to a rock concert with his kids. And then he was heading on a mini-vacation with his wife.

In fact, Gawande had suggested to our mutual friend that I should email him again, later that month, when he would have more time in his schedule for chatting.

At that moment, I realized two things:

First, I was clearly doing something wrong because I hadn't taken a day off in nine months; in fact, I was growing worried that, given a choice between their father and the babysitter, my kids would pick the sitter.

Second, and more important, there were people out there who knew how to be more productive. I just had to convince them to share their secrets with me.

● ● ●

This book is the result of my investigations into how productivity works, and my effort to understand why some people and companies are so much more productive than everyone else.

Since I contacted Gawande four years ago, I've sought out neurologists, businesspeople, government leaders, psychologists, and other productivity experts. I've spoken to the filmmakers behind Disney's *Frozen,* and learned how they made one of the most successful movies in history under crushing time pressures—and narrowly averted disaster—by fostering a certain kind of creative tension within their ranks. I talked to data scientists at Google and writers from the early seasons of *Saturday Night Live* who said both organizations were successful, in part, because they abided by a similar set of unwritten rules regarding mutual support and risk taking. I interviewed FBI agents who solved a kidnapping through agile management and a culture influenced by an old auto plant in Fremont, California. I roamed the halls of Cincinnati's public schools and saw how an initiative to improve education transformed students' lives by, paradoxically, making information *more difficult* to absorb.

As I spoke to people—poker players, airline pilots, military generals, executives, cognitive scientists—a handful of key insights began to emerge. I noticed that people kept mentioning the same concepts over and over. I came to believe a small number of ideas were at the core of why some people and companies get so much done.

This book, then, explores the eight ideas that seem most important in expanding productivity. One chapter, for example, examines how a feeling of control can generate motivation, and how the military turns directionless teenagers into marines by teaching them choices that are "biased toward action." Another chapter looks at why we can maintain focus by constructing mental models—and how one group of pilots told themselves stories that prevented 440 passengers from falling out of the sky.

This book's chapters describe the correct way to set goals—by embracing both big ambitions and small-bore objectives—and why

Israel's leaders became so obsessed with the wrong aspirations in the run-up to the Yom Kippur War. They explore the importance of making decisions by envisioning the future as multiple possibilities rather than fixating on what you hope will happen, and how a woman used that technique to win a national poker championship. They describe how some Silicon Valley companies became giants by building "commitment cultures" that supported employees even when such commitment gets hard.

Connecting these eight ideas is a powerful underlying principle: Productivity isn't about working more or sweating harder. It's not simply a product of spending longer hours at your desk or making bigger sacrifices.

Rather, productivity is about making certain choices in certain ways. The way we choose to see ourselves and frame daily decisions; the stories we tell ourselves, and the easy goals we ignore; the sense of community we build among teammates; the creative cultures we establish as leaders: These are the things that separate the merely busy from the genuinely productive.

We now exist in a world where we can communicate with coworkers at any hour, access vital documents over smartphones, learn any fact within seconds, and have almost any product delivered to our doorstep within twenty-four hours. Companies can design gadgets in California, collect orders from customers in Barcelona, email blueprints to Shenzhen, and track deliveries from anywhere on earth. Parents can auto-sync the family's schedules, pay bills online while lying in bed, and locate the kids' phones one minute after curfew. We are living through an economic and social revolution that is as profound, in many ways, as the agrarian and industrial revolutions of previous eras.

These advances in communications and technology are supposed to make our lives easier. Instead, they often seem to fill our days with more work and stress.

In part, that's because we've been paying attention to the wrong innovations. We've been staring at the tools of productivity—the gadgets and apps and complicated filing systems for keeping track of various to-do lists—rather than the lessons those technologies are trying to teach us.

There are some people, however, who have figured out how to master this changing world. There are some companies that have discovered how to find advantages amid these rapid shifts.

We now know how productivity really functions. We know which choices matter most and bring success within closer reach. We know how to set goals that make the audacious achievable; how to reframe situations so that instead of seeing problems, we notice hidden opportunities; how to open our minds to new, creative connections; and how to learn faster by slowing down the data that is speeding past us.

This is a book about how to recognize the choices that fuel true productivity. It is a guide to the science, techniques, and opportunities that have changed lives. There are people who have learned how to succeed with less effort. There are companies that create amazing things with less waste. There are leaders who transform the people around them.

This is a book about how to become smarter, faster, and better at everything you do.

MOTIVATION

Reimagining Boot Camp, Nursing Home Rebellions, and the Locus of Control

The trip was intended as a celebration, a twenty-nine-day tour of South America that would take Robert, who had just turned sixty, and his wife, Viola, first to Brazil, then over the Andes into Bolivia and Peru. Their itinerary included tours of Incan ruins, a boat trip on Lake Titicaca, the occasional craft market, and a bit of birding.

That much relaxation, Robert had joked with friends before leaving, seemed unsafe. He was already anticipating the fortune he would spend on calls to his secretary. Over the previous half century Robert Philippe had built a small gas station into an auto parts empire in rural Louisiana and had made himself into a Bayou mogul through hard work, charisma, and hustle. In addition to the auto-parts business, he also owned a chemical company, a paper supplier, various swaths of land, and a real estate firm. And now here he was, entering his seventh decade, and his wife had convinced him to spend a month in a bunch of countries where, he suspected, it would be awfully difficult to find a TV showing the LSU–Ole Miss game.

Robert liked to say there wasn't a dirt road or back alley along the Gulf Coast he hadn't driven at least once to drum up business. As Philippe Incorporated had grown, Robert had become famous for dragging big-city businessmen from New Orleans and Atlanta out to ramshackle bars and forbidding them from leaving until the ribs were picked clean and bottles sucked dry. Then, while everyone nursed painful hangovers the next morning, Robert would convince them to sign deals worth millions. Bartenders always knew to fill his glass with club soda while serving the bigwigs cocktails. Robert hadn't touched booze in years.

He was a member of the Knights of Columbus and the chamber of commerce, past president of the Louisiana Association of Wholesalers and the Greater Baton Rouge Port Commission, the chairman of his local bank, and a loyal donor to whichever political party was more inclined to endorse his business permits that day. "You never met a man who loved working so much," his daughter, Roxann, told me.

Robert and Viola had been looking forward to this South American trip. But when they stepped off the plane in La Paz, midway through the monthlong tour, Robert started acting oddly. He staggered through the airport and had to sit down to catch his breath at the baggage claim. When a group of children approached him to ask for coins, Robert threw change at their feet and laughed. In the bus to the hotel, Robert started a loud, rambling monologue about various countries he had visited and the relative attractiveness of the women who lived there. Maybe it was the altitude. At twelve thousand feet, La Paz is one of the highest cities in the world.

Once they were unpacked, Viola urged Robert to nap. He wasn't interested, he said. He wanted to go out. For the next hour, he marched through town buying trinkets and exploding in a rage whenever locals didn't understand English. He eventually agreed to return to the hotel and fell asleep, but woke repeatedly during the night to vomit. The next morning, he said he felt faint but became angry when Viola suggested he rest. He spent the third day in bed. On day four, Viola decided enough was enough and cut the vacation short.

Back home in Louisiana, Robert seemed to improve. His disorientation faded and he stopped saying strange things. His wife and children, however, were still worried. Robert was lethargic and refused to leave the house unless prodded. Viola had expected him to rush into the office upon their return, but after four days he hadn't so much as checked in with his secretary. When Viola reminded him that deer hunting season was approaching and he'd need to get a license, Robert said he thought he'd skip it this year. She phoned a doctor. Soon, they were driving to the Ochsner Clinic in New Orleans.

The chief of neurology, Dr. Richard Strub, put Robert through a battery of tests. Vital signs were normal. Blood work showed nothing unusual. No indication of infection, diabetes, heart attack, or stroke. Robert demonstrated understanding of that day's newspaper and could clearly recall his childhood. He could interpret a short story. The Revised Wechsler Adult Intelligence Scale showed a normal IQ.

"Can you describe your business to me?" Dr. Strub asked.

Robert explained how his company was organized and the details of a few contracts they had recently won.

"Your wife says you're behaving differently," Dr. Strub said.

"Yeah," Robert replied. "I don't seem to have as much get-up-and-go as I used to."

"It didn't seem to bother him," Dr. Strub later told me. "He told me about the personality changes very matter of fact, like he was describing the weather."

Except for the sudden apathy, Dr. Strub couldn't find evidence of illness or injury. He suggested to Viola they wait a few weeks to see if Robert's disposition improved. When they returned a month later, however, there had been no change. Robert wasn't interested in seeing old friends, his wife said. He didn't read anymore. Previously, it had been infuriating to watch television with him because he would flip from channel to channel, looking for a more exciting show. Now, he just stared at the screen, indifferent to what was on. She had finally convinced him to go into the office, but his secretary said he spent hours at his desk gazing into space.

"Are you unhappy or depressed?" Dr. Strub asked.

"No," Robert said. "I feel good."

"Can you tell me how you spent yesterday?"

Robert described a day of watching television.

"You know, your wife tells me your employees are concerned because they don't see you around the office much," said Dr. Strub.

"I guess I'm more interested in other things now," Robert replied.

"Like what?"

"Oh, I don't know," Robert said, and then went silent and stared at the wall.

Dr. Strub prescribed various medications—drugs to combat hormonal imbalances and attention disorders—but none seemed to make a difference. People suffering from depression will say they are unhappy and describe hopeless thoughts. Robert, however, said he was satisfied with life. He admitted his personality change was odd, but it didn't upset him.

Dr. Strub administered an MRI, which allowed him to collect images from inside Robert's cranium. Deep inside his skull, near the center of Robert's head, he saw a small shadow, evidence that burst vessels had caused a tiny amount of blood to pool temporarily inside a part of Robert's brain known as the striatum. Such injuries, in rare cases, can cause brain damage or mood swings. But except for the listlessness, there was little in Robert's behavior to suggest that he was suffering any neurological disability.

A year later, Dr. Strub submitted an article to the *Archives of Neu-*

rology. Robert's "behavior change was characterized by apathy and lack of motivation," he wrote. "He has given up his hobbies and fails to make timely decisions in his work. He knows what actions are required in his business, yet he procrastinates and leaves details unattended. Depression is not present." The cause of this passivity, Dr. Strub suggested, was the slight damage in his brain, which had possibly been triggered by Bolivia's altitude. Even that, however, was uncertain. "It is possible that the hemorrhages are coincidental and that the high altitude played no physiologic role."

It was an interesting but ultimately inconclusive case, Dr. Strub wrote.

● ● ●

Over the next two decades, a handful of other studies appeared in medical journals. There was the sixty-year-old professor who experienced a rapid "decrease in interest." He had been an expert in his field with a fierce work ethic. Then, one day, he simply stopped. "I just lack spirit, energy," he told his physician. "I have no go. I must force myself to get up in the morning."

There was a nineteen-year-old woman who had fallen briefly unconscious after a carbon monoxide leak and then seemed to lose motivation for the most basic tasks. She would sit in one position all day unless forced to move. Her father learned he couldn't leave her alone, as a neurologist wrote, when she "was found by her parents with heavy sunburns on the beach at the very same place where she laid down several hours before, under an umbrella: intense inertia had prevented her from changing her position with that of the shadow while the sun had turned around."

There was a retired police officer who began waking up "late in the morning, would not wash unless urged to do so, but meekly complied as soon as his wife asked him to. Then he would sit in his armchair, from which he would not move." There was a middle-aged man who was stung by a wasp and, not long after, lost the desire to interact with his wife, children, and business associates.

In the late 1980s, a French neurologist in Marseille named Michel Habib heard about a few of these cases, became intrigued, and started searching archives and journals for similar stories. The studies he found were rare but consistent: A relative would bring a patient in for an examination, complaining of a sudden change in behavior and passivity. Doctors would find nothing medically wrong. The patients scored normally when tested for mental illness. They had moderate to high IQs and appeared physically healthy. None of them said they felt depressed or complained about their apathy.

Habib began contacting the physicians treating these patients and asked them to collect MRIs. He then discovered another commonality: All the apathetic individuals had tiny pinpricks of burst vessels in their striatum, the same place where Robert had a small shadow inside his skull.

The striatum serves as a kind of central dispatch for the brain, relaying commands from areas like the prefrontal cortex, where decisions are made, to an older part of our neurology, the basal ganglia, where movement and emotions emerge. Neurologists believe the striatum helps translate decisions into action and plays an important role in regulating our moods. The damage from the burst vessels inside the apathetic patients' striata was small—too small, some of Habib's colleagues said, to explain their behavior changes. Beyond those pinpricks, however, Habib could find nothing else to explain why their motivation had disappeared.

Neurologists have long been interested in striatal injuries because the striatum is involved in Parkinson's disease. But whereas Parkinson's often causes tremors, a loss of physical control, and depression, the patients Habib studied only seemed to lose their drive. "Parkinsonians have trouble initiating movement," Habib told me.

"But the apathetic patients had no problems with motion. It's just that they had no desire to move." The nineteen-year-old woman who couldn't be left alone at the beach, for example, was able to clean her room, wash the dishes, fold the laundry, and follow recipes when instructed to do so by her mother. However, if she *wasn't* asked to help, she wouldn't move all day. When her mother inquired what she wanted for dinner, the woman said she had no preferences.

When examined by doctors, Habib wrote, the apathetic sixty-year-old professor would "stay motionless and speechless during endless periods, sitting in front of the examiner, waiting for the first question." When asked to describe his work, he could discuss complicated ideas and quote papers from memory. Then he would lapse back into silence until another question was posed.

None of the patients Habib studied responded to medications, and none seemed to improve with counseling. "Patients demonstrate a more or less total indifference to life events that would normally provoke an emotional response, positive or negative," Habib wrote.

"It was as if the part of their brain where motivation lives, where élan vital is stored, had completely disappeared," he told me. "There were no negative thoughts, there were no positive thoughts. There were no thoughts at all. They hadn't become less intelligent or less aware of the world. Their old personalities were still inside, but there was a total absence of drive or momentum. Their motivation was completely gone."

II.

The room where the experiment was conducted at the University of Pittsburgh was painted a cheery yellow and contained an fMRI machine, a computer monitor, and a smiling researcher who looked too young to have a PhD. All participants in the study were welcomed into the room, asked to remove their jewelry and any metal from their pockets, and then told to lie on a plastic table that slid into the fMRI.

Once lying down, they could see a computer screen. The researcher explained that a number between one and nine was going to appear on the monitor. Before that number appeared, participants had to guess if it was going to be higher or lower than five by pressing various buttons. There would be multiple rounds of guessing, the researcher said. There was no skill involved in this game, he explained. No abilities were being tested. And though he didn't mention this to the participants, the researcher thought this was one of the most boring games in existence. In fact, he had explicitly designed it that way.

The truth was, the researcher, Mauricio Delgado, didn't care if participants guessed right or wrong. Rather, he was interested in understanding which parts of their brains became active as they played an intensely dull game. As they made their guesses, the fMRI was recording the activity inside their skulls. Delgado wanted to identify where the neurological sensations of excitement and anticipation— where motivation—originated. Delgado told participants they could quit whenever they wanted. Yet he knew, from prior experience, that people would make guess after guess, sometimes for hours, as they waited to see if they had guessed wrong or right.

Each participant lay inside the machine and watched the screen intently. They hit buttons and made predictions. Some cheered when they won or moaned when they lost. Delgado, monitoring the activity inside of their heads, saw that people's striata—that central dispatch—lit up with activity whenever participants played, regardless of the outcome. This kind of striatal activity, Delgado knew, was associated with emotional reactions—in particular, with feelings of expectation and excitement.

As Delgado was finishing one session, a participant asked if he could continue playing on his own, at home.

"I don't think that's possible," Delgado told him, explaining that the game only existed on his computer. Besides, he said, letting the man in on a secret, the experiment was rigged. To make sure the game was consistent from person to person, Delgado had programmed the computer so that everyone won the first round, lost the second, won the third, lost the fourth, and so on, in a predetermined pattern. The outcome had been determined ahead of time. It was like betting on a two-headed quarter.

"That's okay," the man replied. "I don't mind. I just like to play."

"It was odd," Delgado told me later. "There's no reason he should have wanted to continue playing once he knew it was rigged. I mean, where's the fun in a rigged game? Your choices have no impact. But it took me five minutes to convince him he didn't want to take the game home."

For days afterward, Delgado kept thinking about that man. Why had this game interested him so much? For that matter, why had it entertained so many other participants? The experiment's data had helped Delgado identify which parts of people's brains became active as they played a guessing game, but the data didn't explain *why* they were motivated to play in the first place.

So a few years later, Delgado set up another experiment. A new set of participants was recruited. Like before, there was a guessing game. This time, however, there was a key difference: Half the time, participants were allowed to make their own guesses; the rest of the time, the computer guessed for them.

As people began playing, Delgado watched the activity in their striata. This time, when people were allowed to make their own choices, their brains lit up just like in the previous experiment. They showed the neurological equivalents of anticipation and excitement. But during those rounds when participants didn't have any control over their guesses, when the computer made a choice for them, people's striata went essentially silent. It was as if their brains became uninterested in the exercise. There was "robust activity in the caudate nucleus only when subjects" were permitted to guess, Delgado and his colleagues later wrote. "The anticipation of choice itself was associated with increased activity in corticostriatal regions, particularly the ventral striatum, involved in affective and motivational processes."

What's more, when Delgado asked participants about their perceptions of the game afterward, they said they enjoyed themselves much more when they were in control of their choices. They cared whether they won or lost. When the computer was in charge, they said, the experiment felt like an assignment. They got bored and wanted it to end.

That didn't make sense to Delgado. The odds of winning or losing were exactly the same regardless of whether the participant or the computer was in control. Allowing someone to make a guess, rather than waiting for a computer to make a guess for them, shouldn't have made any real difference in the experience of the game. People's neurological reactions *should* have been the same either way. But, somehow, allowing people to make choices transformed the game. Instead of being a chore, the experiment became a challenge. Participants were more motivated to play simply because they believed they were in control.

III.

In recent decades, as the economy has shifted and large companies promising lifelong employment have given way to freelance jobs and migratory careers, understanding motivation has become increasingly important. In 1980, more than 90 percent of the American workforce reported to a boss. Today more than a third of working Americans are freelancers, contractors, or in otherwise transitory positions. The workers who have succeeded in this new economy are those who know how to decide for themselves how to spend their time and allocate their energy. They understand how to set goals, prioritize tasks, and make choices about which projects to pursue. People who know how to self-motivate, according to studies, earn more money than their peers, report higher levels of happiness, and say they are more satisfied with their families, jobs, and lives.

Self-help books and leadership manuals often portray self-motivation as a static feature of our personality or the outcome of a neurological calculus in which we subconsciously compare efforts versus rewards. But scientists say motivation is more complicated than that. Motivation is more like a skill, akin to reading or writing, that can be learned and honed. Scientists have found that people can get better at self-motivation if they practice the right way. The trick, researchers say, is realizing that a prerequisite to motivation is believing we have authority over our actions and surroundings. To motivate ourselves, we must feel like we are in control.

"The need for control is a biological imperative," a group of Columbia University psychologists wrote in the journal *Trends in Cognitive Sciences* in 2010. When people believe they are in control, they tend to work harder and push themselves more. They are, on average, more confident and overcome setbacks faster. People who believe they have authority over themselves often live longer than their peers. This instinct for control is so central to how our brains develop that infants, once they learn to feed themselves, will resist adults' attempts at control even if submission is more likely to get food into their mouths.

One way to prove to ourselves that we are in control is by making decisions. "Each choice—no matter how small—reinforces the perception of control and self-efficacy," the Columbia researchers wrote. Even if making a decision delivers no benefit, people still want the freedom to choose. "Animals and humans demonstrate a preference for choice over non-choice, even when that choice confers no additional reward," Delgado noted in a paper published in the journal *Psychological Science* in 2011.

From these insights, a theory of motivation has emerged: The first step in creating drive is giving people opportunities to make choices that provide them with a sense of autonomy and self-determination. In experiments, people are more motivated to complete difficult tasks when those chores are presented as decisions rather than commands. That's one of the reasons why your cable company asks all those questions when you sign up for service. If they ask if you prefer a paperless bill to an itemized statement, or the ultra package versus the platinum lineup, or HBO to Showtime, you're more likely to be motivated to pay the bill each month. As long as we feel a sense of control, we're more willing to play along.

"You know when you're stuck in traffic on the freeway and you see an exit approaching, and you want to take it even though you know it'll probably take longer to get home?" said Delgado. "That's our brains getting excited by the possibility of taking control. You won't get home any faster, but it *feels* better because you feel like you're in charge."

This is a useful lesson for anyone hoping to motivate themselves or others, because it suggests an easy method for triggering the will to act: Find a choice, almost any choice, that allows you to exert control. If you are struggling to answer a tedious stream of emails, decide to reply to one from the middle of your inbox. If you're trying to start an assignment, write the conclusion first, or start by making the graphics, or do whatever's most interesting to you. To find the motivation to confront an unpleasant employee, choose where the meeting is going to occur. To start the next sales call, decide what question you'll ask first.

Motivation is triggered by making choices that demonstrate to ourselves that we are in control. The specific choice we make matters less than the assertion of control. It's this feeling of self-determination that gets us going. That's why Delgado's participants were willing to play again and again when they felt like they were in charge.

Which is not to say that motivation is, therefore, always easy. In fact, sometimes simply making a choice isn't enough. Occasionally, to really self-motivate, we need something more.

IV.

After Eric Quintanilla signed his name to the form that officially made him a U.S. Marine, the recruiter shook his hand, looked him in the eye, and said he had made the right choice.

"It's the only one I see for myself, sir," Quintanilla replied. He had meant the words to sound bold and confident, but his voice quavered when he spoke and his hand was so sweaty that both of them wiped their palms on their pants afterward.

Quintanilla was twenty-three years old. Five years earlier, he had graduated from high school in a small town an hour south of Chicago. He had thought about going away to college, but he wasn't certain what to study, wasn't positive what he wanted to do afterward—wasn't sure about much, to be honest. So he enrolled at a local community college and got an associate's degree in general studies. He had hoped it would help him get a job at a cellphone store in the mall. "I filled out, I don't know, like ten applications," Quintanilla said. "But I never heard back from anyone."

He found part-time work at a hobby supply shop and occasionally drove an ice truck when the regular guy was sick or on vacation. At night, he played World of Warcraft. This wasn't how Quintanilla had envisioned his life. He was ready for something better. He decided to propose to the girl he had been dating since high school. The wedding was fantastic. Afterward, though, he was still in the same place. And then his wife got pregnant. He tried the cellphone stores once more and scored an interview. He rehearsed with his wife the night before his appointment.

"Honey," she told him, "you have to give them a reason to hire you. Just tell them what you're excited about."

The next day, when the store manager asked him why he wanted to sell T-Mobile phones, Quintanilla froze. "I don't know," he said. It was the truth. He had no idea.

A few weeks later, Quintanilla went to a party and saw one of his former classmates, freshly home from basic training and twenty pounds lighter, with bulging muscles and a newfound sense of confidence. He was telling jokes and hitting on girls. Maybe, Quintanilla said to his wife the next morning, he should consider the Marines. She didn't like the idea, and neither did his mom, but Quintanilla couldn't think of anything else to do. He sat down one night at the kitchen table, drew a line down the center of a piece of paper, wrote "Marine Corps" on the left side and tried to fill the right with other options. The only thing he could come up with was "Get promoted at the hobby store."

Five months later, he arrived at the San Diego Marine Corps Recruit Depot in the middle of the night, shuffled into a room alongside eighty other young men, had his head shaved, his blood type tested, his clothes replaced with fatigues, and embarked on a new life.

The thirteen-week boot camp Quintanilla entered in 2010 was a relatively new experiment in the Corps' 235-year-old quest to manufacture the perfect marine. For most of its history, the service's training program had focused on molding rowdy teenagers into disciplined troops. But fifteen years before Quintanilla's enlistment, a fifty-three-year-old general named Charles C. Krulak had been promoted to commandant, the Marines' top position. Krulak believed basic training needed to change. "We were seeing much weaker applicants," he told me. "A lot of these kids didn't just need discipline, they needed a mental makeover. They'd never belonged to a sports team, they'd never had a real job, they'd never *done* anything. They didn't even have the vocabulary for ambition. They'd followed instructions their whole life."

This was a problem, because the Corps increasingly needed troops who could make independent decisions. Marines—as they will happily tell you—are different from soldiers and sailors. "We're the first to arrive and the last to leave," Krulak said. "We need extreme self-starters." In today's world, that means the Corps requires men and women capable of fighting in places such as Somalia and Baghdad, where rules and tactics change unpredictably and marines often have to decide—on their own and in real time—the best course of action.

"I began spending time with psychologists and psychiatrists, trying to figure out, how do we do a better job teaching these recruits to think for themselves?" Krulak said. "We had great recruits coming in, but they didn't have any sense of direction or drive. All they knew was doing the bare minimum. It was like working with a bunch of wet socks. Marines can't be wet socks."

Krulak began reviewing studies on how to teach self-motivation, and became particularly intrigued by research, conducted by the Corps years earlier, showing that the most successful marines were those with a strong "internal locus of control"—a belief they could influence their destiny through the choices they made.

Locus of control has been a major topic of study within psychology since the 1950s. Researchers have found that people with an internal locus of control tend to praise or blame themselves for success or failure, rather than assigning responsibility to things outside their influence. A student with a strong internal locus of control, for instance, will attribute good grades to hard work, rather than natural smarts. A salesman with an internal locus of control will blame a lost sale on his own lack of hustle, rather than bad fortune.

"Internal locus of control has been linked with academic success, higher self-motivation and social maturity, lower incidences of stress and depression, and longer life span," a team of psychologists wrote in the journal *Problems and Perspectives in Management* in 2012. People with an internal locus of control tend to earn more money, have more friends, stay married longer, and report greater professional success and satisfaction.

In contrast, having an *external* locus of control—believing that your life is primarily influenced by events outside your control—"is correlated with higher levels of stress, [often] because an individual perceives the situation as beyond his or her coping abilities," the team of psychologists wrote.

Studies show that someone's locus of control can be influenced through training and feedback. One experiment conducted in 1998, for example, presented 128 fifth graders with a series of difficult puzzles. Afterward, each student was told they had scored very well. Half of them were also told, "You must have worked hard at these problems." Telling fifth graders they have worked hard has been shown to activate their *internal* locus of control, because hard work is something we decide to do. Complimenting students for hard work reinforces their belief that they have control over themselves and their surroundings.

The other half of the students were also informed they had scored well, and then told, "You must be really smart at these problems." Complimenting students on their intelligence activates an *external* locus of control. Most fifth graders don't believe they can choose how smart they are. In general, young kids think that intelligence is an innate capacity, so telling young people they are smart reinforces their belief that success or failure is based on factors *outside* of their control.

Then all the students were invited to work on three more puzzles of varying difficulty.

The students who had been praised for their intelligence— who had been primed to think in terms of things they could *not* influence—were much more likely to focus on the easier puzzles during the second round of play, even though they had been complimented for being smart. They were less motivated to push themselves. They later said the experiment wasn't much fun.

In contrast, students who had been praised for their hard work—who were encouraged to frame the experience in terms of self-determination—went to the hard puzzles. They worked longer and scored better. They later said they had a great time.

"Internal locus of control is a learned skill," Carol Dweck, the Stanford psychologist who helped conduct that study, told me. "Most of us learn it early in life. But some people's sense of self-determination gets suppressed by how they grow up, or experiences they've had, and they forget how much influence they can have on their own lives.

"That's when training is helpful, because if you put people in situations where they can *practice* feeling in control, where that internal locus of control is reawakened, then people can start building habits that make them feel like they're in charge of their own lives—and the more they feel that way, the more they really are in control of themselves."

For Krulak, studies like this seemed to hold the key to teaching recruits self-motivation. If he could redesign basic training to force trainees to take control of their own choices, that impulse might become more automatic, he hoped. "Today we call it teaching 'a bias toward action,'" Krulak told me. "The idea is that once recruits have taken control of a few situations, they start to learn how good it feels.

"We never tell anyone they're a natural-born leader. 'Natural born' means it's outside your control," Krulak said. "Instead, we teach them that leadership is learned, it's the product of effort. We push recruits to experience that thrill of taking control, of feeling the rush of being in charge. Once we get them addicted to that, they're hooked."

For Quintanilla, this tutorial started as soon as he arrived. Initially, there were long days of forced marches, endless sit-ups and push-ups, and tedious rifle drills. Instructors screamed at him constantly. ("We've got an image to uphold," Krulak told me.) But alongside those exercises, Quintanilla also confronted a steady stream of situations that forced him to make decisions and take control.

In his fourth week of training, for instance, Quintanilla's platoon was told to clean the mess hall. The recruits had no idea how. They didn't know where the cleaning supplies were located or how the industrial dishwasher worked. Lunch had just ended and they weren't sure if they were supposed to wrap the leftovers or throw them away. Whenever someone approached a drill instructor for advice, all he received was a scowl. So the platoon began making choices. The potato salad got tossed, the leftover hamburgers went into the fridge, and the dishwasher was loaded with so much detergent that suds soon covered the floor. It took three and a half hours, including the time spent mopping up the bubbles, for the platoon to finish cleaning the mess hall. They mistakenly threw away edible food, accidentally turned off the ice cream freezer, and somehow managed to misplace two dozen forks. When they were done, however, their drill instructor approached the smallest, shyest member of the platoon and said he had noticed how the recruit had asserted himself when a decision was needed on where to put the ketchup. In truth, it was pretty obvious where the ketchup should have gone. There was a huge set of shelves containing nothing but ketchup bottles. But the shy recruit beamed as he was praised.

"I hand out a number of compliments, and all of them are designed to be unexpected," said Sergeant Dennis Joy, a thoroughly intimidating drill instructor who showed me around the Recruit Depot one day. "You'll never get rewarded for doing what's easy for you. If you're an athlete, I'll never compliment you on a good run. Only the small guy gets congratulated for running fast. Only the shy guy gets recognized for stepping into a leadership role. We praise people for doing things that are hard. That's how they learn to believe they can do them."

● ● ●

The centerpiece of Krulak's redesigned basic training was the Crucible, a grueling three-day challenge at the end of boot camp. Quintanilla was terrified of the Crucible. He and his bunkmates whispered about it at night. There were rumors and wild conjectures. Someone said a recruit had lost a limb midway through the course the previous year. Quintanilla's Crucible began on a Tuesday morning when his platoon was woken at two A.M. and told they should prepare themselves to march, crawl, and climb across fifty miles of obstacle courses. Each person carried thirty pounds of gear. They had only two meals apiece to last fifty-four hours. At most, they could hope for just a few hours of sleep. Injuries were expected. Anyone who stopped moving or lagged too far behind, they were told, would be dropped from the Corps.

Midway through the Crucible, the recruits encountered a task called Sergeant Timmerman's Tank. "The enemy has chemically contaminated this area," a drill instructor shouted, pointing to a pit the size of a football field. "You must cross it while wearing full gear and gas masks. If a recruit touches the ground, you have failed and must start over. If you spend longer than sixty minutes in the pit, you have failed and must start over. You must obey your team leader. I repeat: You may not proceed without a direct verbal order from the team leader. You must *hear* a command before you act, otherwise you have failed and must start over."

Quintanilla's team formed a circle and began using a technique they had learned in basic training.

"What's our objective?" one recruit said.

"To cross the pit," someone replied.

"How do we use the boards?" another recruit said, pointing to planks with ropes attached.

"We could lay them end to end," someone answered. The team leader issued a verbal order and the circle broke up to test this idea along the border of the pit. They stood on one board while hauling the other forward. No one could keep their balance. The circle reformed. "How do we use the ropes?" a recruit asked.

"To lift the planks," another said. He suggested standing on both boards simultaneously and using the ropes to lift each piece in tandem, as if on skis.

Everyone put on their gas masks and stood on the boards with the leader at the front. "Left!" he shouted as recruits pulled one of the planks slightly forward. "Right!" They began shuffling across the pit. After ten minutes, however, it was clear this wasn't working. Some people were lifting too quickly, others were pushing the boards too far. And because they were all wearing gas masks, it was impossible to hear the leader's commands. They had already gone too far to turn around—but at this rate, crossing would take hours. Recruits began yelling at each other to stop.

The leader ordered a pause. He turned to the man behind him. "Watch my shoulders," he yelled through his gas mask. The leader shrugged his left shoulder, and then his right. By watching the rhythm of the leader, the recruit behind him could coordinate how to lift the boards. The only problem with this idea was that it violated one of the ground rules. Recruits had been told they could not act until they heard a verbal command from their team leader. But with their gas masks on, no one could really hear anything. However, there was no other way to proceed. So the team leader began shrugging and swinging his arms while screaming orders. No one caught on at first, so he began yelling one of the songs they had learned on long marches. The recruit behind him could make out enough of what he was singing to join in. His neighbor did the same. Eventually, they were all singing and shrugging and swinging in tandem. They crossed the field in twenty-eight minutes.

"Technically, we could send them back to start over because each person didn't hear a direct verbal command from the team leader," a drill sergeant later told me. "But that's the point of the exercise: We know you can't hear anything with the gas masks on. The only way to get across the pit is to figure out some workaround. We're trying to teach them that you can't just obey orders. You have to take control and figure things out for yourself."

Twenty-four hours and another dozen obstacles later, Quintanilla's platoon gathered at the base of the Crucible's final challenge, a long, steep hill they called the Grim Reaper. "You don't *have* to help each other during the Reaper," Krulak said. "I've seen that happen before. Recruits fall down, and they don't have buddies, so they get left behind."

Quintanilla had been marching for two days by this point. He had slept less than four hours. His face was numb and his hands were covered with blisters and cuts from carrying water-filled drums across obstacles. "There were guys throwing up at the Reaper," he told me. "One person had his arm in a sling." As the group began walking up the mountain, recruits kept stumbling. They were all so exhausted they moved as if in slow motion, hardly making any progress. So they began linking up, arm in arm, to prevent one another from sliding down the incline.

"Why are you doing this?" Quintanilla's pack buddy wheezed at him, lapsing into a call-and-response they had practiced on hikes. When things are at their most miserable, their drill instructors had said, they should ask each other questions that begin with "why."

"To become a Marine and build a better life for my family," Quintanilla said.

His wife had given birth a week earlier to a daughter, Zoey. He had been allowed to speak to her for a total of five minutes by telephone after the delivery. It was his only contact with the outside world in almost two months. If he finished the Crucible, he would see his wife and new child.

If you can link something hard to a choice you care about, it makes the task easier, Quintanilla's drill instructors had told him. That's why they asked each other questions starting with "why." Make a chore into a *meaningful* decision, and self-motivation will emerge.

The platoon summited the last peak as the sun crested, and staggered to a clearing with a flagpole. Everyone went still. They were finally done. The Crucible was over. A drill instructor walked through their formation, pausing before each man to place the service's insignia, the Eagle, Globe, and Anchor, in their hands. They were officially marines.

"You think boot camp is going to be all screaming and fighting," Quintanilla told me. "But it's not. It's not like that at all. It's more about learning how to make yourself do things you thought you couldn't do. It's really emotional, actually."

Basic training, like the Marine Corps career itself, offers few material rewards. A Marine's starting salary is $17,616 a year. However, the Corps has one of the highest career satisfaction rates. The training the Corps provides to roughly forty thousand recruits each year has transformed the lives of millions of people who, like Quintanilla, had no idea how to generate the motivation and self-direction needed to take control of their lives. Since Krulak's reforms, the Corps' retention of new recruits and the performance scores of new marines have both increased by more than 20 percent. Surveys indicate that the average recruit's internal locus of control increases significantly during basic training. Delgado's experiments were a start to understanding motivation. The Marines complement those insights by helping us understand how to teach drive to people who aren't practiced in self-determination: If you give people an opportunity to feel a sense of control and let them practice making choices, they can learn to exert willpower. Once people know how to make self-directed choices into a habit, motivation becomes more automatic.

Moreover, to teach ourselves to self-motivate more easily, we need to learn to see our choices not just as expressions of control but also as affirmations of our values and goals. That's the reason recruits ask each other "why"—because it shows them how to link small tasks to larger aspirations.

The significance of this insight can be seen in a series of studies conducted in nursing homes in the 1990s. Researchers were studying why some seniors thrived inside such facilities, while others experienced rapid physical and mental declines. A critical difference, the researchers determined, was that the seniors who flourished made choices that rebelled against the rigid schedules, set menus, and strict rules that the nursing homes tried to force upon them.

Some researchers referred to such residents as "subversives," because so many of their decisions manifested as small rebellions against the status quo. One group at a Santa Fe nursing home, for instance, started every meal by trading food items among themselves in order to construct meals of their own design rather than placidly accept what had been served to them. One resident told a researcher that he always gave his cake away because, even though he liked cake, he would "rather eat a second-class meal that I have chosen."

A group of residents at a nursing home in Little Rock violated the institution's rules by moving furniture around to personalize their bedrooms. Because wardrobes were attached to the walls, they used a crowbar—appropriated from a tool closet—to wrench their dressers free. In response, an administrator called a meeting and said there was no need to undertake independent redecorations; if the residents needed help, the staff would provide it. The residents informed the administrator that they didn't want any assistance, didn't need permission, and intended to continue doing whatever they damn well pleased.

These small acts of defiance were, in the grand scheme of things, relatively minor. But they were psychologically powerful because the subversives saw the rebellions as evidence that they were still in control of their own lives. The subversives walked, on average, about twice as much as other nursing home residents. They ate about a third more. They were better at complying with doctors' orders, taking their medications, visiting the gym, and maintaining relationships with family and friends. These residents had arrived at the nursing homes with just as many health problems as their peers, but once inside, they lived longer, reported higher levels of happiness, and were far more active and intellectually engaged.

"It's the difference between making decisions that prove to yourself that you're still in charge of your life, versus falling into a mindset where you're just waiting to die," said Rosalie Kane, a gerontologist at the University of Minnesota. "It doesn't really matter if you eat cake or not. But if you refuse to eat *their* cake, you're demonstrating to yourself that you're still in charge." The subversives thrived because they knew how to take control, the same way that Quintanilla's troop learned to subversively cross a pit during the Crucible by deciding, on their own, how to interpret the rules.

The choices that are most powerful in generating motivation, in other words, are decisions that do two things: They convince us we're in control *and* they endow our actions with larger meaning. Choosing to climb a mountain can become an articulation of love for a daughter. Deciding to stage a nursing home insurrection can become proof that you're still alive. An internal locus of control emerges when we develop a mental habit of transforming chores into *meaningful choices*, when we assert that we have authority over our lives.

Quintanilla finished boot camp in 2010 and served in the Corps for three years. He then left. He was finally ready, he felt, for real life. He got another job, but the lack of camaraderie among his colleagues was disappointing. No one seemed motivated to excel. So in 2015, he reenlisted. "I missed that constant reminder that I can do anything," he told me. "I missed people pushing me to choose a better me."

V.

Viola Philippe, the wife of the onetime auto parts tycoon of Louisiana, was something of an expert on motivation herself before she and Robert flew to South America. She had been born with albinism—her body did not produce the enzyme tyrosinase, critical in the production of melanin—and as a result, her skin, hair, and eyes contained no pigment, and her eyesight was poor. She was legally blind, and could read only by putting her face very close to a page and using a magnifying glass. "You have never met a more determined person, though," her daughter, Roxann, told me. "She could do anything."

When Viola was a girl, the school district had tried to put her into remedial classes despite the fact that it was her eyes, not her brain, that had problems. But she refused to leave the classroom where her friends sat. She stayed in that room until administrators relented. After she graduated, she went to Louisiana State University and told the school she expected them to provide someone to read textbooks to her aloud. The school complied. During her sophomore year, she met Robert, who soon dropped out to start washing and greasing cars for a local Ford dealer. He encouraged her to quit school, as well. She politely declined and got her degree. They were married in December 1950, four months after she graduated.

They had six children in rapid succession, and while Robert built his empire, Viola ran the household. There were morning meetings and charts showing what each child had to accomplish each day. There were Friday night check-ins, during which everyone laid out their goals for the coming week. "They were like two peas in a pod, both totally driven," said Roxann. "Mom refused to let her disabilities stop her. I think that's why it was so hard for her when Dad changed."

When Robert's apathy took hold, Viola initially focused her energy on caring for him. She hired nurses to help him exercise, and worked with his brother to form a committee to oversee and then sell off Robert's companies. After a while, however, she ran out of things to do. She had married a bon vivant, a man so full of life that it was hard to go to the grocery store because he constantly stopped to chat with everyone. Now Robert sat in a chair in front of the television all day. Viola was miserable. "He didn't speak to me," she told a courtroom when the family sued for insurance money they felt owed because of Robert's neurological injuries. "He wasn't—it didn't seem like he was interested in anything I did. You know, I would fix his meals and I was more or less a caretaker. I guess you would call me a caretaker."

For a few years, she felt sorry for herself. Then she became angry. Then busy. If Robert wasn't going to show any motivation to reclaim his life, then she would force him to get moving again. She would make him engage. She began by asking him ceaseless questions. When she made lunch, she would pepper him with choices. Sandwich or soup? Lettuce or tomato? Ham or turkey? What about mayonnaise? Ice water or juice? She didn't really mean anything by it at first. She was just frustrated and wanted to make him speak.

But then, after a few months of harassing him, Viola found that whenever Robert was pressured into making decisions, he seemed to come out of his shell a little bit. He would banter with her for a few moments, or tell her about a program he had been watching. One night, after she had forced him to make a dozen choices about what he was going to eat and which table they should sit at and what music to listen to, he began talking at length, reminding her of a funny story from after they had gotten married, when they had locked themselves out of the house in a rainstorm. He told the story in an offhand way, and chuckled as he recalled trying to jimmy a window. It was the first time Viola had heard him laugh in years. For a few minutes, it was like the old Robert was back. Then he faced the TV and went silent again.

Viola continued her campaign, and over time, more and more of the old Robert emerged. Viola congratulated, cajoled, and rewarded him whenever he seemed, for a moment, like his former self. When he went back to Dr. Strub, the neurologist in New Orleans, for his annual checkup seven years after the trip to South America, the doctor could see the difference. "He was saying hello to the nurses, and asking them about their kids," Dr. Strub said. "He would initiate conversations with me, ask about my hobbies. He had opinions on the route they should drive to get home. It was stuff you wouldn't have noticed with anyone else, but with him, it was like someone was turning on the lights again."

As neurologists have studied how motivation functions within our brains, they've become increasingly convinced that people like Robert don't lose their drive because they've lost the capacity for self-motivation. Rather, their apathy is due to an *emotional* dysfunction. One of the things Habib, the French researcher, noticed about all the people he studied was that they shared an odd emotional detachment. One apathetic woman told him she had hardly reacted when her father died. A man said he hadn't felt the urge to hug his wife or children since the passivity had taken hold. When Habib asked patients if they felt sad about how much their lives had changed, they all said no. They didn't feel anything.

Neurologists have suggested that this emotional numbness is why some people feel no motivation. Among Habib's patients, the injuries in their striata prevented them from feeling the sense of reward that comes from taking control. Their motivation went dormant because they had forgotten how good it feels to make a choice. In other situations, it's that people have never learned what it feels like to be self-determined, because they have grown up in a neighborhood that seems to offer so few choices or they have forgotten the rewards of autonomy since they've moved into a nursing home.

This theory suggests how we can help ourselves and others strengthen our internal locus of control. We should reward initiative, congratulate people for self-motivation, celebrate when an infant wants to feed herself. We should applaud a child who shows defiant, self-righteous stubbornness and reward a student who finds a way to get things done by working around the rules.

This is easier in theory, of course, than practice. We all applaud self-motivation until a toddler won't put on his shoes, an aged parent is ripping a dresser out of the wall, or a teenager ignores the rules. But that's how an internal locus of control becomes stronger. That's how our mind learns and remembers how good it *feels* to be in control. And unless we practice self-determination and give ourselves emotional rewards for subversive assertiveness, our capacity for self-motivation can fade.

What's more, we need to prove to ourselves that our choices are meaningful. When we start a new task, or confront an unpleasant chore, we should take a moment to ask ourselves "why." Why are we forcing ourselves to climb up this hill? Why are we pushing ourselves to walk away from the television? Why is it so important to return that email or deal with a coworker whose requests seem so unimportant?

Once we start asking why, those small tasks become pieces of a larger constellation of meaningful projects, goals, and values. We start to recognize how small chores can have outsized emotional rewards, because they prove to ourselves that we are making meaningful choices, that we are genuinely in control of our own lives. That's when self-motivation flourishes: when we realize that replying to an email or helping a coworker, on its own, might be relatively unimportant. But it is part of a bigger project that we believe in, that we want to achieve, that we have *chosen* to do. Self-motivation, in other words, is a choice we make because it is part of something bigger and more emotionally rewarding than the immediate task that needs doing.

In 2010, twenty-two years after her South American vacation with Robert, Viola was diagnosed with ovarian cancer. It took two years for the disease to consume her. At every step, Robert was there, helping her out of bed in the morning and reminding her to take her medications at night. He asked her questions to distract her from the pain and fed her when she became feeble. When Viola finally passed, Robert sat by her empty bed for days. His children, worried he was slipping back into apathy, suggested another visit with the neurologist in New Orleans. Perhaps the doctor would recommend something to forestall his listlessness from returning.

No, Robert replied. It wasn't apathy keeping him indoors. He just needed some time to reflect on sixty-two years of marriage. Viola had helped Robert build a life—and then, when everything had slipped from his grasp, she helped him rebuild it again. He just wanted to honor that by pausing for a few days, he told his kids. A week later, he left the house and came over for brunch. Afterward, he babysat his grandchildren. Robert passed away twenty-four months later, in 2014. He was active, his obituary noted, until the end.

2

TEAMS

Psychological Safety at Google and *Saturday Night Live*

Julia Rozovsky was twenty-five years old and uncertain what to do with her life when she decided it was time for a change. She was a Tufts graduate with a bachelor's degree in math and economics who had previously worked at a consulting firm, which she found unfulfilling. Then she had become a researcher for two professors at Harvard, which was fun but not a long-term career.

Maybe, she thought, she belonged in a big corporation. Or perhaps she ought to become an academic. Or maybe she should join a tech start-up. It was all very confusing to her. So she picked the option that meant she didn't have to decide: She applied to business schools, and was accepted to start at the Yale School of Management in 2010.

She showed up in New Haven ready to bond with her classmates and, like all new students, was assigned to a study group. This group, she figured, would be an important part of her education.

They would become close friends and learn together, debate important issues, and discover, with each other's help, who they were meant to be.

Study groups are a rite of passage at most MBA programs, a way for students to practice working in teams. At Yale, "each study group shares the same class schedule and collaborates on each group assignment," one of the school's websites explained. "Study groups have been carefully constructed to bring together students with diverse backgrounds, both professionally and culturally." Each day during lunch or after dinner, Julia and the four other members of her study group would gather to discuss homework and compare spreadsheets, strategize for upcoming exams, and trade lecture notes. Truth be told, her group wasn't all that diverse. Two of them had been management consultants, like Julia. Another had worked at a start-up. They were all smart and curious and outgoing. Their similarities, she hoped, would make it easy for them to bond. "There are lots of people who say some of their best business school friends come from their study groups," said Julia. "But it wasn't like that for me."

Almost from the start, study group felt like a daily dose of stress. "I never felt completely relaxed," she told me. "I always felt like I had to prove myself." Dynamics quickly emerged that put her on edge. Everyone wanted to show they were leaders, and so when teachers issued study group assignments, there were subtle tussles over who was in charge. "People would try to show authority by speaking louder, or talking over each other," Julia said. When it came to divvying up tasks for projects, one group member would sometimes preemptively assign roles, and then the others would critique those assignments, and then someone else would claim authority over some part of the project, and then everyone else would rush to grab their own piece. "Maybe it was my own insecurities, but I always felt like I had to be careful not to make mistakes around them," said

Julia. "People were critical of each other, but they would play it off like they were making a joke, and so the group was kind of passive-aggressive.

"I was looking forward to making friends with my group," she said. "It really bummed me out that we didn't gel."

So Julia started looking for other groups to join, other ways to connect with classmates. One person mentioned that some students were putting together a team to participate in "case competitions," in which business school students proposed innovative solutions to real-world business problems. Teams would receive a case study, spend a few weeks writing a business plan, and then submit it to high-profile executives and professors who picked the winner. Companies sponsored these contests and there were cash prizes as well as, sometimes, jobs that came out of the competitions. Julia signed up.

Yale hosted about a dozen different case competition teams. The one Julia joined included a former army officer, a think tank researcher, the director of a health education nonprofit, and a refugee program manager. Unlike her study group, everyone was from different backgrounds. From the start, though, they all clicked. Each time a new case arrived, the team would gather in the library and dive into action, spending hours brainstorming options, assigning research duties, and divvying up writing assignments. Then they would meet again and again and again. "One of the best cases we did was about Yale itself," Julia said. "There had always been a student-run snack store, but the university was taking over food sales, and so the business school sponsored a contest to overhaul the shop.

"We met every night for a week. I thought we should fill the shop with nap pods, and someone else said it should become a game room, and there was also some kind of clothing swap idea. We had lots of crazy ideas." No one ever shot down a suggestion, not even the nap pods. Julia's study group, as part of their class assignments, had also engaged in a fair amount of brainstorming, "but if I had ever mentioned something like a nap pod, somebody would have rolled their eyes and come up with fifteen reasons why it was a dumb idea. And it *was* a dumb idea. But my case team loved it. We always loved each other's dumb ideas. We spent an hour figuring out how nap pods could make money by selling accessories like earplugs."

Eventually, Julia's case team settled on the idea of converting the student shop into a micro-gym with a handful of exercise classes and a few workout machines. They spent weeks researching pricing models and contacting equipment manufacturers. They won the competition and the micro-gym exists today. That same year, Julia's case team spent another month studying ways for a chain of eco-friendly convenience stores to expand into North Carolina. "We must have analyzed two dozen plans," she said. "A lot of them turned out not to make any sense." When the team traveled to Portland, Oregon, to present their final suggestion—a slow-growth approach that emphasized the chain's healthy food options—they placed first in the nation.

Julia's study group dissolved sometime in her second semester after one person, and then another, and then everyone stopped showing up. The case competition team grew as new students asked if they could join. The core group of five teammates, including Julia, remained involved the entire time they were at Yale. Today, these people are some of her closest friends. They attend one another's weddings and visit each other when traveling. They call each other for career advice and pass along job leads.

It always struck Julia as odd that those two teams *felt* so different. Her study group felt stressful because everyone was always jousting for leadership and critiquing each other's ideas. Her case competition team felt exciting because everyone was so supportive and enthusiastic. Both groups, however, were composed of basically the same kinds of people. They were all bright, and everyone was friendly outside of the team settings. There was no reason why the dynamic inside Julia's study group needed to become so competitive, while the culture of the case team was so easygoing.

"I couldn't figure out why things had turned out so different," Julia told me. "It didn't seem like it had to happen that way."

● ● ●

After graduation, Julia went to work at Google and joined its People Analytics group, which was tasked with studying nearly every aspect of how employees spent their time. What she was supposed to do with her life, it turned out, was use data to figure out why people behave in certain ways.

For six years running, Google had been ranked by *Fortune* as one of America's top workplaces. The company's executives believed that was because, even as it had grown to fifty-three thousand employees, Google had devoted enormous resources to studying workers' happiness and productivity. The People Analytics group, part of Google's human resources division, helped examine if employees were satisfied with their bosses and coworkers, whether they felt overworked, intellectually challenged, and fairly paid, whether their work-life balance was actually balancing out, as well as hundreds of other variables. The division helped with hiring and firing decisions, and its analysts provided insights into who should be promoted and who, perhaps, had risen too fast. In the years before Julia joined the group, People Analytics had determined that Google needed to interview a job applicant only four times to predict, with 86 percent confidence, if they would be a good hire. The division had successfully pushed to increase paid maternity leave from twelve to eighteen weeks because computer models indicated that would reduce the frequency of new mothers quitting by 50 percent. At the most basic level, the division's goal was to make life at Google a little bit better and a lot more productive. With enough data, People Analytics believed, almost any behavioral puzzle could be solved.

People Analytics' biggest undertaking in recent years had been a study—code-named Project Oxygen before it was revealed—that examined why some managers were more effective than others. Ultimately, researchers had identified eight critical management skills.* "Oxygen was a huge success for us," said Abeer Dubey, a People Analytics manager. "It helped clarify what differentiated good managers from everyone else and how we could help people improve." The project was so useful, in fact, that at about the same time Julia was hired, Google began another massive effort, this one code-named Project Aristotle.

Dubey and his colleagues had noticed that many Google employees, in company surveys, had consistently mentioned the importance of their teams. "Googlers would say things like 'I have a great manager, but my team has never clicked' or 'My manager isn't fantastic, but the team is so strong it doesn't matter,'" said Dubey. "And that was kind of eye opening, because Project Oxygen had looked at leadership, but it hadn't focused on how teams function, or if there's an optimal mix of different kinds of people or backgrounds." Dubey and his colleagues wanted to figure out how to build the perfect team. Julia became one of the effort's researchers.

The project started with a sweeping review of academic literature. Some scientists had found that teams functioned best when they contained a concentration of people with similar levels of extroversion and introversion, while others had found that a balance of personalities was key. There were studies about the importance of teammates having similar tastes and hobbies, and others lauding diversity within groups. Some research suggested that teams needed people who like to collaborate; others said groups were more successful when individuals had healthy rivalries. The literature, in other words, was all over the place.

---

* Project Oxygen found that a good manager (1) is a good coach; (2) empowers and does not micromanage; (3) expresses interest and concern in subordinates' success and well-being; (4) is results oriented; (5) listens and shares information; (6) helps with career development; (7) has a clear vision and strategy; (8) has key technical skills.

So Project Aristotle spent more than 150 hours asking Google employees what *they* thought made a team effective. "We learned that teams are somewhat in the eye of the beholder," said Dubey. "One group might appear like it's working really well from the outside, but, inside, everyone is miserable." Eventually, they established criteria for measuring teams' effectiveness based on external factors, such as whether a group hit their sales targets, as well as internal variables, such as how productive team members felt. Then the Aristotle group began measuring everything they could. Researchers examined how often teammates socialized outside of work and how members divided up tasks. They drew complicated diagrams to show teams' overlapping memberships, and then compared those against statistics of which groups had exceeded their department's goals. They studied how long teams stuck together and if gender balance had an impact on effectiveness.

No matter how they arranged the data, though, it was almost impossible to find patterns—or any evidence that a team's composition was correlated with its success. "We looked at 180 teams from all over the company," said Dubey. "We had *lots* of data, but there was nothing showing that a mix of specific personality types or skills or backgrounds made any difference. The 'who' part of the equation didn't seem to matter."

Some productive Google teams, for instance, were composed of friends who played sports together outside of work. Others were made up of people who were basically strangers away from the conference room. Some groups preferred strong managers. Others wanted a flatter structure. Most confounding of all, sometimes two teams would have nearly identical compositions, with overlapping memberships, but radically different levels of effectiveness. "At Google, we're good at finding patterns," said Dubey. "There weren't strong patterns here."

So Project Aristotle turned to a different approach. There was a second body of academic research that focused on what are known as "group norms." "Any group, over time, develops collective norms about appropriate behavior," a team of psychologists had written in the *Sociology of Sport Journal*. Norms are the traditions, behavioral standards, and unwritten rules that govern how we function. When a team comes to an unspoken consensus that avoiding disagreement is more valuable than debate, that's a norm asserting itself. If a team develops a culture that encourages differences of opinion and spurns groupthink, that's another norm holding sway. Team members might behave certain ways as individuals—they may chafe against authority or prefer working independently—but often, inside a group, there's a set of norms that override those preferences and encourage deference to the team.

The Project Aristotle researchers went back to their data and analyzed it again, this time looking for norms. They found that some teams consistently allowed people to interrupt one another. Others enforced taking conversational turns. Some teams celebrated birthdays and began each meeting with a few minutes of informal chitchat. Others got right to business. There were teams that contained extroverts who hewed to the group's sedate norms whenever they assembled, and others where introverts came out of their shells as soon as meetings began.

And some norms, the data indicated, consistently correlated with high team effectiveness. One engineer, for instance, told the researchers that his team leader "is direct and straightforward, which creates a safe space for you to take risks. . . . She also takes the time to ask how we are, figure out how she can help you and support you." That was one of the most effective groups inside Google.

Alternately, another engineer told the researchers that his "team leader has poor emotional control. He panics over small issues and keeps trying to grab control. I would hate to be driving with him in the passenger seat, because he would keep trying to grab the steering wheel and crash the car." That team did not perform well.

Most of all, though, employees talked about how various teams *felt.* "And that made a lot of sense to me, maybe because of my experiences at Yale," Julia said. "I'd been on some teams that left me feeling totally exhausted and others where I got so much energy from the group."

There is strong evidence that group norms play a critical role in shaping the emotional experience of participating in a team. Research by psychologists from Yale, Harvard, Berkeley, the University of Oregon, and elsewhere indicate that norms determine whether we feel safe or threatened, enervated or excited, and motivated or discouraged by our teammates. Julia's study group at Yale, for instance, felt draining because the norms—the tussles over leadership, the pressure to constantly demonstrate expertise, the tendency to critique—had put her on guard. In contrast, the norms of her case competition team—enthusiasm for one another's ideas, withholding criticisms, encouraging people to take a leadership role or hang back as they wanted—allowed everyone to be friendly and unconstrained. Coordination was easy.

Group norms, the researchers on Project Aristotle concluded, were the answer to improving Google's teams. "The data finally started making sense," said Dubey. "We had to manage the *how* of teams, not the *who.*"

The question, however, was which norms mattered most. Google's research had identified dozens of norms that seemed important—and, sometimes, the norms of one effective team contradicted the norms of another, equally successful group. Was it better to let everyone speak as much as they wanted, or should strong leaders end meandering debates? Was it more effective for people to openly disagree with one another, or should conflicts be downplayed? Which norms were most crucial?

II.

In 1991, a first-year PhD student named Amy Edmondson began visiting hospital wards, intending to show that good teamwork and good medicine went hand in hand. But the data kept saying she was wrong.

Edmondson was studying organizational behavior at Harvard. A professor had asked her to help with a study of medical mistakes, and so Edmondson, on the prowl for a dissertation topic, started visiting recovery rooms, talking to nurses, and paging through error reports from two Boston hospitals. In one cardiac ward, she discovered that a nurse had accidentally given a patient an IV of lidocaine, an anesthetic, rather than heparin, a blood thinner. In an orthopedic ward, a patient was given amphetamines rather than aspirin. "You would be shocked at how many mistakes occur every day," Edmondson told me. "Not because of incompetence, but because hospitals are really complicated places and there's usually a large team—as many as two dozen nurses and techs and doctors—who might be involved in each patient's care. That's a lot of opportunities for something to slip through the cracks."

Some parts of the hospitals Edmondson visited seemed more accident prone than others. The orthopedic ward, for instance, reported an average of one error every three weeks; the cardiac ward, on the other hand, reported a mistake almost every other day. Edmondson also found that the various departments had very different cultures. In the cardiac ward nurses were chatty and informal; they gossiped in the hallways and had pictures of their kids on the walls. In orthopedics, people were more sedate. Nurse managers wore business suits rather than scrubs and asked everyone to keep the public areas free of personal items and clutter. Perhaps, Edmondson thought, she could study the various teams' cultures and see if they correlated with error rates.

She and a colleague created a survey to measure team cohesion on various wards. She asked nurses to describe how frequently their team leader set clear goals and whether teammates discussed conflicts openly or avoided tense conversations. She measured the satisfaction, happiness, and self-motivation of different groups and hired a research assistant to observe the wards for two months.

"I figured it would be pretty straightforward," Edmondson told me. "The units with the strongest sense of teamwork would have the lowest error rates." Except, when she tabulated her data, Edmondson found exactly the opposite. The wards with the strongest team cohesion had far *more* errors. She checked the data again. It didn't make any sense. Why would strong teams make more mistakes?

Confused, Edmondson decided to look at these nurses' responses, question by question, alongside the error rates to see if any explanations emerged. Edmondson had included one survey question that inquired specifically about the personal risks associated with making errors. She asked people to agree or disagree with the statement: "If you make a mistake in this unit, it is held against you." Once she compared the data from that question with error incidence, she realized what was going on. It wasn't that wards with strong teams were making more mistakes. Rather, it was that nurses who belonged to strong teams felt more comfortable *reporting* their mistakes. The data indicated that one particular norm—whether people were punished for missteps—influenced if they were honest after they screwed up.

Some leaders "have established a climate of openness that facilitates discussion of error, which is likely to be an important influence on detected error rates," Edmondson wrote in *The Journal of Applied Behavioral Science* in 1996. What particularly surprised her, however, was how complicated things got the closer she looked: it wasn't simply that strong teams encouraged open communication and weak teams discouraged it. In fact, while some strong teams emboldened people to admit their mistakes, other, equally strong teams made it hard for nurses to speak up. What made the difference wasn't team cohesion—rather, it was the culture each team established. In one ward with a strong team, for instance, nurses were overseen by "a hands-on manager who actively invites questions and concerns. . . . In an interview, the nurse manager explains that a 'certain level of error will occur' so a 'nonpunitive environment' is essential to deal with this error productively," Edmondson wrote. "There is an unspoken rule here to help each other and check each other," a nurse told Edmondson's assistant. "People feel more willing to admit to errors here, because the nurse manager goes to bat for you."

In another ward with a team that, at first glance, seemed equally strong, a nurse said that when she admitted hurting a patient while drawing blood, the nurse manager "made her feel like she was on trial." Another said doctors "bite your head off if you make a mistake." Yet measurements of group cohesion on this ward were still very high. A nurse told the research assistant that the ward "prides itself on being clean, neat and having an appearance of professionalism." The nurse manager for the ward dressed in business suits and when she delivered criticism, she considerately offered her critiques behind closed doors. The staff said they appreciated the manager's professionalism, were proud of their department, and felt a strong sense of unity. To Edmondson, the team seemed like they genuinely liked and respected one another. But they also admitted that the unit's culture sometimes made it hard to confess making a mistake.

It wasn't the strength of the team that determined how many errors were reported—rather, it was one specific norm.

When Edmondson started working on her dissertation, she visited technology companies and factory floors, and asked people about the unwritten rules that shaped how their teammates behaved. "People would say things like, 'This is one of the best teams I've ever been on, because I don't have to wear a work face here,' or 'We aren't afraid to share crazy ideas,'" Edmondson told me. On those teams, norms of enthusiasm and support had taken hold and everyone felt empowered to voice opinions and take risks. "And other teams would tell me, 'My group is really dedicated to each other and so I try not to go outside my department without checking with my supervisor first' or 'We're all in this together, so I don't like to bring up an idea unless I know it will work.'" Within those teams, a norm of loyalty held sway—and it undermined people's willingness to make suggestions or take chances.

Both enthusiasm and loyalty are admirable norms. It wasn't clear to managers that they would have such different impacts on people's behaviors. And yet they did. In that setting, enthusiastic norms made teams better. Loyalty norms made them less effective. "Managers never intend to create unhealthy norms," Edmondson said. "Sometimes, though, they make choices that seem logical, like encouraging people to flesh out their ideas before presenting them, that ultimately undermine a team's ability to work together."

As her research continued, Edmondson found a handful of good norms that seemed to be consistently associated with higher productivity. On the best teams, for instance, leaders encouraged people to speak up; teammates felt like they could expose their vulnerabilities to one another; people said they could suggest ideas without fear of retribution; the culture discouraged people from making harsh judgments. As Edmondson's list of good norms grew, she began to notice that everything shared a common attribute: They were all behaviors that created a sense of togetherness while also encouraging people to take a chance.

"We call it 'psychological safety,'" she said. Psychological safety is a "shared belief, held by members of a team, that the group is a safe place for taking risks." It is "a sense of confidence that the team will not embarrass, reject, or punish someone for speaking up," Edmondson wrote in a 1999 paper. "It describes a team climate characterized by interpersonal trust and mutual respect in which people are comfortable being themselves."

Julia and her Google colleagues found Edmondson's papers as they were researching norms. The idea of psychological safety, they felt, captured everything their data indicated was important to Google's teams. The norms that Google's surveys said were most effective—allowing others to fail without repercussions, respecting divergent opinions, feeling free to question others' choices but also trusting that people aren't trying to undermine you—were all aspects of feeling psychologically safe at work. "It was clear to us that this idea of psychological safety was pointing to which norms were most important," said Julia. "But it wasn't clear how to teach those inside Google. People here are really busy. We needed clear guidelines for creating psychological safety without losing the capacity for dissent and debate that's critical to how Google functions." In other words, how do you convince people to feel safe while also encouraging them to be willing to disagree?

"For a long time, that was the million-dollar question," Edmondson told me. "We knew it was important for teammates to be open with each other. We knew it was important for people to feel like they can speak up if something's wrong. But those are also the behaviors that can set people at odds. We didn't know why some groups could clash and still have psychological safety while others would hit a period of conflict and everything would fall apart."

III.

On the first day of auditions for the television show that became known as *Saturday Night Live,* the actors showed up, one after another, hour after hour, until it felt like it would never stop. There were two women who played midwestern housewives preparing for the annual meteorological disaster ("Can I borrow your centerpiece for the tornado this year?") and a singer with an original composition named "I Am Dog" lampooning the women's liberation anthem "I Am Woman." A roller-skating impressionist and an obscure musician named Meat Loaf took the stage around lunchtime. The actor Morgan Freeman and the comic Larry David were on the call sheet, as were four jugglers and five mimes. To the exhausted observers watching the auditions, it felt as if every vaudeville act and stand-up comedian between Boston and Washington, D.C., had shown up.

Which is the way the show's thirty-year-old creator, Lorne Michaels, wanted it. Over the previous nine months, Michaels had traveled from Bangor to San Diego, watching hundreds of comedy club shows. He talked to writers from television and radio programs and every magazine with a humor page. His goal, he later said, was to see "every single funny person in North America."

By noon on the second day of auditions, tryouts were running late when a man burst through the doors, leapt onto the stage, and demanded the producers' attention. He had a trim mustache and wore a three-piece suit. He carried a folded umbrella and an attaché. "I've been waiting out there for three hours and I'm not going to wait anymore!" he shouted. "I'm going to miss my plane!" He marched across the stage. "That's it! You've had your chance! Good day!" Then he stormed out.

"What the hell was that?" one producer asked.

"Oh, that was just Danny Aykroyd," said Michaels. They had known each other in Toronto, where Aykroyd was a student in Michaels's improv class. "He's probably going to do the show," Michaels said.

Over the next month, as Michaels chose the rest of the cast, the same thing happened again and again: Instead of picking from among the hundreds of people he auditioned, Michaels hired comedians he already knew or who had been recommended by friends. Michaels knew Aykroyd from Canada, and Aykroyd, in turn, was enthusiastic about a guy named John Belushi he had met in Chicago. Belushi initially said he'd never appear on television because it was a crass medium, but he recommended a castmate from the *National Lampoon Show* named Gilda Radner (who Michaels, it turned out, had already hired; they knew each other from *Godspell*). The *National Lampoon Show* was affiliated with *National Lampoon* magazine, which was founded by the writer Michael O'Donoghue, who lived with another comedy writer named Anne Beatts.

All of these people created the first season of *Saturday Night Live*. Howard Shore, the show's music director, had gone to summer camp with Michaels. Neil Levy, the show's talent coordinator, was Michaels's cousin. Michaels had met Chevy Chase while standing in a line in Hollywood to see *Monty Python and the Holy Grail*. Tom Schiller, another writer, knew Michaels because they had gone to Joshua Tree to eat hallucinogenic mushrooms together, and Schiller's father, a Hollywood writer, had taken Michaels under his wing early in the young man's career.

The original cast and writers of *Saturday Night Live* hailed largely from Canada, Chicago, and Los Angeles and all moved to New York in 1975. "Manhattan was a show business wasteland then," said Marilyn Suzanne Miller, a writer whom Michaels hired after they collaborated on a Lily Tomlin special in L.A. "It was like Lorne had deposited us on Mars."

When most of the staff got to New York, they didn't know anyone except one another. Many considered themselves anticapitalist or antiwar activists—or, at least, they were fond of the recreational drugs these activists enjoyed—and now they were riding elevators with a bunch of suits at 30 Rockefeller Center, where the show's studio was being built. "We were all like twenty-one or twenty-two years old. We didn't have any money, or any clue what we were doing, so we spent all of our time trying to make each other laugh," Schiller told me. "We'd eat every meal together. We'd go to the same bars each night. We were terrified that if we separated, one of us might get lost and never be heard from again."

In subsequent years, as *Saturday Night Live* became one of the most popular and longest-running programs in television history, a kind of mythology emerged. "In the early days of *SNL*," the journalist Malcolm Gladwell wrote in 2002, "everyone knew everyone and everyone was always in everyone else's business, and that fact goes a long way toward explaining the extraordinary chemistry among the show's cast." There are books filled with stories of John Belushi breaking into castmates' apartments to make spaghetti late at night, or setting their guest bedrooms on fire with carelessly handled joints, or writers gluing one another's furniture to the ceilings, or prank calling one another's offices, or ordering thirty pizzas to the news division and then dressing up like security guards so they could infiltrate the lower floors, steal the pizza, and leave the journalists with the bill. There are flowcharts detailing who from *SNL* slept with whom. (They tend to get complicated, because Michaels was married to writer Rosie Shuster, who eventually ended up with Dan Aykroyd, who had dated Gilda Radner, who everyone suspected was in love with writer Alan Zweibel, who later wrote a book explaining they *were* in love, but nothing ever happened and, besides, Radner later married a member of the *SNL* band. "It was the 1970s," Miller told me. "Sex was what you did.")

Saturday Night Live has been held up as a model of great team dynamics. It is cited in college textbooks as an example of what groups can achieve when the right conditions are in place and a team intensely bonds.

The group that created *Saturday Night Live* came together so successfully, this theory goes, because a communal culture replaced individual needs. There were shared experiences ("We were all the kids who didn't get to sit at the popular table in high school," Beatts told me); common social networks ("Lorne was a cult leader," said writer Bruce McCall. "As long as you had a Moonie-like devotion to the group, you were fine."); and group needs trumped individual egos ("I don't mean this in a bad way, but we were Guyana on the seventeenth floor," said Zweibel. "It was a stalag.").

But this theory becomes considerably more complicated when you speak to the people on the original *Saturday Night Live* team. It's true those writers and actors spent enormous amounts of time together and developed a strong sense of unity—but not because of forced intimacy or shared history or because they particularly liked each other. In fact, the group norms at *Saturday Night Live* created as many tensions as strengths. "There was a tremendous amount of competitiveness and infighting," said Beatts. "We were so young, and no one knew how to control themselves. We fought all the time."

One night in the writers' room, Beatts made a joke that they were lucky Hitler had killed six million Jews because, otherwise, no one would have found an apartment in New York City. "Marilyn Miller didn't speak to me for two weeks," she said. "Marilyn was completely uptight about jokes about Hitler. I think she hated me at that point. We would glare at each other for hours." There were jealousies and rivalries, battles for Michaels's affection, competition for airtime. "You wanted your sketch to go on, which meant someone else's would have to get cut," said Beatts. "If you were succeeding, someone else was failing."

Even the closest relationships, such as between Alan Zweibel and Gilda Radner, were fraught. "Gilda and I came up with this character, Roseanne Roseannadanna, and on Friday I would go into the office and stay up all night writing the script, like eight or nine pages," said Zweibel. "Then Gilda would arrive midmorning, totally refreshed, and take a red pen and start crossing shit out, like she was some kind of schoolmarm, and I would get pissed. So I would go back to my office and rework everything, and she would do it again. By the time the show went on, we usually weren't speaking to each other. I once stopped writing sketches for her for three weeks. I purposely saved my best stuff for other people."

Furthermore, it's not entirely true that members of the *SNL* team enjoyed spending time together. Garrett Morris, the show's only black actor, felt like an outcast and planned to quit as soon as he had enough money. Jane Curtin would escape to her home and husband as soon as the show was done for the week. People would form allegiances, and then get into fights, and then form counter-allegiances. "Everyone was in these cliques that were constantly shifting," said

Bruce McCall, who came aboard as a writer for the show's second season. "It was a pretty dismal place."

In some ways it's remarkable the *Saturday Night Live* team gelled at all. Michaels, it turned out, had chosen everyone precisely because of their disparate tastes. Zweibel was a specialist in borscht belt one-liners. Michael O'Donoghue wrote dark, bitter satires about such topics as the assassination of JFK. (When a distraught secretary told O'Donoghue that Elvis had died, he replied, "Smart career move.") Tom Schiller aspired to direct art films. And everyone could become scathing critics when their sensibilities clashed. "Great, Garrett," O'Donoghue once said when he read a script the actor had spent weeks writing. Then he dropped it into a trash can. "Real good."

"Comedy writers carry a lot of anger," said Schiller. "We were vicious to each other. If you thought something was funny and no one else did, it could be brutal."

So why, given all the tensions and infighting, did the *Saturday Night Live* creators become such an effective, productive team? The answer isn't that they spent so much time together, or that the show's norms put the needs of the group above individual egos.

Rather, the *SNL* team clicked because, surprisingly, they all felt safe enough around one another to keep pitching new jokes and ideas. The writers and actors worked amid norms that made everyone feel like they could take risks and be honest with one another, even as they were shooting down ideas, undermining one another, and competing for airtime.

"You know that saying, 'There's no *I* in *TEAM*'?" Michaels told me. "My goal was the opposite of that. All I wanted were a bunch of *I*'s. I wanted everyone to hear each other, but no one to disappear into the group."

That's how psychological safety emerged.

●●●

Imagine you have been invited to join one of two teams.

Team A is composed of eight men and two women, all of whom are exceptionally smart and successful. When you watch a video of them working together, you see articulate professionals who take turns speaking and are polite and courteous. At some point, when a question arises, one person—clearly an expert on the topic—speaks at length while everyone else listens. No one interrupts. When another person veers off topic, a colleague gently reminds him of the agenda and steers the conversation back on track. The team is efficient. The meeting ends exactly when scheduled.

Team B is different. It's evenly divided among men and women, some of whom are successful executives, while others are middle managers with little in the way of professional achievements. On a video, you see teammates jumping in and out of a discussion haphazardly. Some ramble at length; others are curt. They interrupt one another so much, it's sometimes hard to follow the conversation. When a team member abruptly changes the topic or loses sight of their point, the rest of the group follows him off the agenda. At the end of the meeting, the meeting doesn't actually end: Everyone sits around and gossips.

Which group would you rather join?

Before you decide, imagine you are given one additional piece of information. When both teams first formed, each member was asked to complete what's known as the "Reading the Mind in the

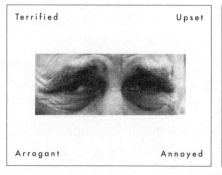

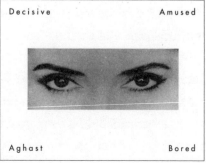

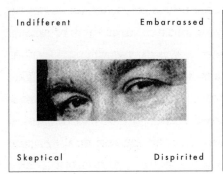

Indifferent	Embarrassed
Skeptical	Dispirited

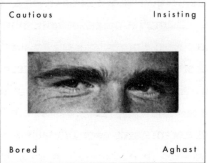

Cautious	Insisting
Bored	Aghast

Eyes" test. They were each shown thirty-six photos of people's eyes and asked to choose which word, among four offered, best described the emotion that person was feeling.*

This test, you are told, measures people's empathy. The members of Team A picked the right emotion, on average, 49 percent of the time. Team B: 58 percent.

Does that change your mind?

In 2008, a group of psychologists from Carnegie Mellon and MIT wondered if they could figure out which kinds of teams were clearly superior. "As research, management, and many other kinds of tasks are increasingly accomplished by groups—both those working face-to-face and 'virtually'—it is becoming even more important to understand the determinants of group performance," the researchers wrote in the journal *Science* in 2010. "Over the last century, psychologists made significant progress in defining and systematically measuring intelligence in individuals. We have used the statistical approach they developed for individual intelligence to systematically measure the intelligence of groups."

Put differently, the researchers wanted to know if there is a collective intelligence that emerges within a team that is distinct from the smarts of any single member.

To accomplish this, the researchers recruited 699 people, divided them into 152 teams, and gave each group a series of assignments

* The correct answers for these photos can be found in the notes on page 309.

that required different kinds of cooperation. Most teams began by spending ten minutes brainstorming possible uses for a brick and received a point for each unique idea. Then they were asked to plan a shopping trip as if they were housemates sharing a single car: Each teammate was given a different list of groceries to buy and a map showing prices at various stores. The only way to maximize the team's score was for each person to sacrifice one item they really wanted in exchange for something that pleased the entire group. Then the teams were told to arrive at a ruling on a disciplinary case in which a college basketball player allegedly bribed his teacher. Some teammates represented the interests of the faculty; others were stand-ins for the athletics department. Points were awarded for reaching a verdict that maximized each group's concerns.

Each of these tasks required full team participation; each demanded different kinds of collaboration. As the researchers observed groups going about the tasks, they saw various dynamics emerge. Some teams came up with dozens of clever uses for the brick, arrived at a verdict that made everyone happy, and easily divvied up the shopping trip. Others kept describing the same uses for the brick in different words; came to verdicts that left some participants feeling alienated; and managed to buy only ice cream and Froot Loops because no one was willing to compromise. What was interesting was that teams that did well on one assignment also seemed to do well on the others. Conversely, teams that failed at one thing seemed to fail at everything.

Some might have hypothesized that the "good teams" were successful because their members were smarter—that group intelligence might be nothing more than the intelligence of the individuals making up the team. But the researchers had tested participants' IQs beforehand and found that individual intelligence didn't correlate with team performance. Putting ten smart people in a room didn't mean they solved problems more intelligently—in fact, those smart people were often outperformed by groups consisting of people who had scored lower on intellect tests, but who still seemed smarter as a group.

Others might have argued that the good teams had more decisive leaders. But the research showed that wasn't right, either.

The researchers eventually concluded that the good teams had succeeded not because of innate qualities of team members, but because of how they treated one another. Put differently, the most successful teams had norms that caused everyone to mesh particularly well.

"We find converging evidence of a general collective intelligence factor that explains a group's performance on a wide variety of tasks," the researchers wrote in their *Science* article. "This kind of collective intelligence is a property of the group itself, not just the individuals in it." It was the norms, not the people, that made teams so smart. The right norms could raise the collective intelligence of mediocre thinkers. The wrong norms could hobble a group made up of people who, on their own, were all exceptionally bright.

But when the researchers reviewed videos of the good teams' interactions, they noticed that not all norms looked alike. "It was striking how different some of them behaved," said Anita Woolley, the study's lead author. "Some teams had a bunch of smart people who figured out how to break up work evenly. Other groups had pretty average members but came up with ways to take advantage of everyone's relative strengths. Some groups had one strong leader. Others were more fluid, and everyone took a leadership role."

There were, however, two behaviors that all the good teams shared.

First, all the members of the good teams spoke in roughly the same proportion, a phenomenon the researchers referred to as "equality in distribution of conversational turn-taking." In some teams, for instance, everyone spoke during each task. In other groups, conversation ebbed from assignment to assignment—but by the end of the day, everyone had spoken roughly the same amount.

"As long as everyone got a chance to talk, the team did well," said Woolley. "But if only one person or a small group spoke all the time, the collective intelligence declined. The conversations didn't need to be equal every minute, but in aggregate, they had to balance out."

Second, the good teams tested as having "high average social sensitivity"—a fancy way of saying that the groups were skilled at intuiting how members felt based on their tone of voice, how people held themselves, and the expressions on their faces.

One of the easiest ways to gauge social sensitivity is to show someone photos of people's eyes and ask them to describe what that person is thinking or feeling—the empathy test described previously. This is a "test of how well the participant can put themselves into the mind of the other person, and 'tune in' to their mental state," wrote the creator of the "Reading the Mind in the Eyes" test, Simon Baron-Cohen of the University of Cambridge. While men, on average, correctly guess the emotion of the person in the photo only 52 percent of the time, women typically guess right 61 percent.

People on the good teams in Woolley's experiment scored above average on the "Reading the Mind in the Eyes" test. They seemed to know when someone was feeling upset or left out. They spent time asking one another what they were thinking about. The good teams also contained more women.

Coming back to the question of which team to join, if you are given a choice between the serious-minded, professional Team A, or the free-flowing, more informal Team B, you should opt for Team B. Team A is smart and filled with effective colleagues. As individuals, they will all be successful. But as a team, they still tend to act like individuals. There's little to suggest that, as a group, they become *collectively* intelligent, because there's little evidence that everyone has an equal voice and that members are sensitive to teammates' emotions and needs.

In contrast, Team B is messier. People speak over one another, they go on tangents, they socialize instead of remaining focused on the agenda. Everyone speaks as much as they need to, though. They feel equally heard and are attuned to one another's body language and expressions. They try to anticipate how one another will react. Team B may not contain as many individual stars, but when that group unites, the sum is much greater than any of its parts.

● ● ●

If you ask the original *Saturday Night Live* team why the show was such a success, they'll talk about Lorne Michaels. There's something about his leadership, they'll say, that made everything come together. He had an ability to make everyone feel heard, to make even the most self-centered actors and writers pay attention to each other. His eye for talent is nearly unrivaled in entertainment over the last forty years.

You'll also find people who say that Michaels is aloof, socially awkward, proud, and jealous, and that when he decides to fire someone, he'll cut them completely adrift. You might not want Michaels as a friend. But as the leader of *Saturday Night Live,* what he's created is extraordinary: one of the longest-running shows in history, built on the talent of egomaniacal comedians who, twenty times a year for four decades, have put their craziness aside just long enough to make a live television program with only a week's preparation.

Michaels himself, still the show's executive producer, says the reason why *Saturday Night Live* has succeeded is because he works hard to force people to become a team. The secret to making that happen, he says, is giving everyone a voice and finding people willing to be sensitive enough to listen to one another.

"Lorne was deliberate about making sure everyone got a chance to pitch their ideas," the writer Marilyn Miller told me. "He would say, 'Do we have pieces for the girls this week?' 'Who hasn't been on in a while?'"

"He has this kind of psychic ability to draw in everyone," said

Alan Zweibel. "I honestly believe that's why the show has existed for forty years. At the top of each script, there's a list of the initials of everyone who worked on that sketch and Lorne has always said he's happiest the more initials he sees."

Michaels is almost ostentatious in his demonstrations of social sensitivity—and he expects the cast and writers to mimic him. During the early years of the show, he was the one who appeared with a soothing word when an exhausted writer was crying in his office. He has been known to interrupt a rehearsal or table read and quietly take an actor aside to ask if they need to talk about something going on in their personal life. Once, when the writer Michael O'Donoghue was inordinately proud of an obscene commercial parody, Michaels ordered it read at eighteen different rehearsals—even though everyone knew the network's censors would never let it on the air.

"I remember walking up to Lorne once and saying, 'Okay, here's my idea, it's a bunch of girls at their first slumber party and they are telling each other how sex works.' And Lorne said, 'Write it up,' just like that, no questions asked. Then he took an index card and put it on the board for the next show." That sketch—which appeared on *Saturday Night Live* on May 8, 1976—became one of the show's most famous pieces. "I was on top of the world," said Miller. "He's got this social ESP. Sometimes he knows exactly what will make you feel like the most important person on earth."

Many of the original actors and writers on *Saturday Night Live* weren't particularly easy to get along with. They freely admit that, even today, they are combative and gossipy and sometimes downright mean. But when they worked together, they were careful with one another's feelings. Michael O'Donoghue might have dropped Garrett Morris's script into a trash can, but he made a point, afterward, to tell Morris he was joking, and when Morris suggested an idea about a depressing children's story, O'Donoghue came up with "The Little Train That Died." ("I know I can! I know I can! Heart attack! Heart attack! Oh, my God, the pain!") The *SNL* team avoided picking fights with one another. ("When I made that Hitler joke, Marilyn wouldn't speak to me," Beatts told me. "But that's the point. She didn't speak. She didn't escalate it into a whole big thing.") People might have criticized one another's ideas, but they were careful about how far they let their critiques go. They disagreed and clashed, but everyone still had a voice at each table read, and despite the sniping and competition, they were oddly protective of one another. "Everyone liked everyone else, or at least worked hard to *pretend* like they liked everyone," said Don Novello, a writer on the show in the 1970s and '80s and the actor who played Father Guido Sarducci. "We genuinely trusted each other, as crazy as that sounds."

For psychological safety to emerge among a group, teammates don't have to be friends. They do, however, need to be socially sensitive and ensure everyone feels heard. "The best tactic for establishing psychological safety is demonstration by a team leader," as Amy Edmondson, who is now a professor at Harvard Business School, told me. "It seems like fairly minor stuff, but when the leader goes out of their way to make someone feel listened to, or starts a meeting by saying 'I might miss something, so I need all of you to watch for my mistakes,' or says 'Jim, you haven't spoken in a while, what do you think?,' that makes a huge difference."

In Edmondson's hospital studies, the teams with the highest levels of psychological safety were also the ones with leaders most likely to model listening and social sensitivity. They invited people to speak up. They talked about their own emotions. They didn't interrupt other people. When someone was concerned or upset, they showed the group that it was okay to intervene. They tried to anticipate how people would react and then worked to accommodate those reactions. This is how teams encourage people to disagree while still being honest with one another and occasionally clashing. This is how psychological safety emerges: by giving everyone an equal voice and encouraging social sensitivity among teammates.

Michaels himself says the job of modeling norms is his most important duty. "Everyone who comes through this show is different, and I have to *show* each of them that I'm treating them different, and show *everyone else* I'm treating them different, if we want to draw the unique brilliance out of everyone," Michaels told me.

"*SNL* only works when we have different writing and performing styles all bumping into and meshing with each other," he said. "That's my job: To protect people's distinct voices, but also to get them to work together. I want to preserve whatever made each person special before they came to the show, but also help everyone be sensitive enough to make the rough edges fit. That's the only way we can do a new show every week without everyone wanting to kill each other as soon as we're done."

IV.

By the summer of 2015, the Google researchers working on Project Aristotle had been collecting surveys, conducting interviews, running regressions, and analyzing statistics for two years. They had scrutinized tens of thousands of pieces of data and had written dozens of software programs to analyze trends. Finally, they were ready to reveal their conclusions to the company's employees.

They scheduled a meeting at the headquarters in Mountain View. Thousands of employees showed up, and many more watched via video stream. Laszlo Bock, the head of the People Operations department at Google, walked onto the stage and thanked everyone for coming. "The biggest thing you should take away from this work is that *how* teams work matters, in a lot of ways, more than *who* is on them," he said.

He had spoken to me before he went onstage. "There's a myth we all carry inside our head," Bock said. "We think we need superstars. But that's not what our research found. You can take a team of average performers, and if you teach them to interact the right way, they'll do things no superstar could ever accomplish. And there's other myths, like sales teams should be run differently than engineering teams, or the best teams need to achieve consensus around everything, or high-performing teams need a high volume of work to stay engaged, or teams need to be physically located together.

"But now we can say those aren't right. The data shows there's a universality to how good teams succeed. It's important that everyone on a team feels like they have a voice, but whether they actually get to vote on things or make decisions turns out not to matter much. Neither does the volume of work or physical co-location. What matters is having a voice and social sensitivity."

Onstage, Bock brought up a series of slides. "What matters are five key norms," he told the audience.

Teams need to believe that their work is important.

Teams need to feel their work is personally meaningful.

Teams need clear goals and defined roles.

Team members need to know they can depend on one another.

But, most important, teams need psychological safety.

To create psychological safety, Bock said, team leaders needed to model the right behaviors. There were Google-designed checklists they could use: Leaders should not interrupt teammates during conversations, because that will establish an interrupting norm. They should demonstrate they are listening by summarizing what people

say after they said it. They should admit what they don't know. They shouldn't end a meeting until all team members have spoken at least once. They should encourage people who are upset to express their frustrations, and encourage teammates to respond in nonjudgmental ways. They should call out intergroup conflicts and resolve them through open discussion.

There were dozens of tactics on the checklist. All of them, however, came back to two general principles: Teams succeed when everyone feels like they can speak up and when members show they are sensitive to how one another feels.

"There are lots of small things a leader can do," Abeer Dubey told me. "In meetings, does the leader cut people off by saying 'Let me ask a question there,' or does she wait until someone is done speaking? How does the leader act when someone's upset? These things are so subtle, but they can have a huge impact. Every team is different, and it's not uncommon in a company like Google for engineers or salespeople to be taught to fight for what they believe in. But you need the right norms to make arguments productive rather than destructive. Otherwise, a team never becomes stronger."

For three months, Project Aristotle traveled from division to division explaining their findings and coaching team leaders. Google's top executives released tools that any team could use to evaluate if members felt psychologically safe and worksheets to help leaders and teammates improve their scores.

"I come from a quantitative background. If I'm going to believe something, you need to give me data to back it up," said Sagnik Nandy, who as chief of Google Analytics Engineering heads one of the company's biggest teams. "So seeing this data has been a game changer for me. Engineers love debugging software because we know we can get 10 percent more efficiency by just making a few tweaks. But we never focus on debugging human interactions. We put great people together and hope it will work, and sometimes it does and sometimes it doesn't, and most of the time we don't know why. Aristotle let us debug our people. It's totally changed how I run meetings. I'm so much more conscious of how I model listening now, or whether I interrupt, or how I encourage everyone to speak."

The project has had an impact on the Aristotle team, as well. "A couple of months ago, we were in a meeting where I made a mistake," Julia Rozovsky told me. "Not a huge mistake, but an embarrassing one, and afterward, I sent out a note explaining what had gone wrong, why it had happened, and what we were doing to resolve it. Right afterward, I got an email back from a team member that just said, 'Ouch.'

"It was like a punch to the gut. I was already upset about making this mistake, and this note totally played on my insecurities. But because of all the work we've done, I pinged the person back and said, 'Nothing like a good *Ouch* to destroy psychological safety in the morning!' And he wrote back and said, 'I'm just testing your resilience.' That could have been the wrong thing to say to someone else, but he knew it was exactly what I needed to hear. With one thirty-second interaction, we diffused the tension.

"It's funny to do a project on team effectiveness while working on a team, because we get to test everything we're learning as we go along. What I've realized is that as long as everyone feels like they can talk and we're really demonstrating that we want to hear each other, you feel like everyone's got your back."

Over the last two decades, the American workplace has become much more team focused. The average worker today might belong to a sales team, as well as a group of unit managers, a special team planning future products, and the team overseeing the holiday party. Executives belong to groups that oversee compensation and strategy and hiring and firing and approving HR policies and figuring out how to cut costs. These teams might meet every day in person or correspond via email or telecommute from all over the world. Teams are important. Within companies and conglomerates, government agencies and schools, teams are now the fundamental unit of self-organization.

And the unwritten rules that make teams succeed or fail, it turns out, are the same from place to place. The way investment bankers coordinate their efforts might seem different from how orthopedic nurses divvy up tasks. And the specific norms, in those different settings, will likely vary. But one thing will remain true if those teams work well: In both places, the groups will feel a sense of psychological safety. They will succeed because teammates feel they can trust each other, and that honest discussion can occur without fear of retribution. Their members will have roughly equal voices. Teammates will show they are sensitive to one another's emotions and needs.

In general, the route to establishing psychological safety begins with the team's leader. So if you are leading a team—be it a group of coworkers or a sports team, a church gathering, or your family dinner table—think about what message your choices send. Are you encouraging equality in speaking, or rewarding the loudest people? Are you modeling listening? Are you demonstrating a sensitivity to what people think and feel, or are you letting decisive leadership be an excuse for not paying as close attention as you should?

There are always good reasons for choosing behaviors that undermine psychological safety. It is often more efficient to cut off debate, to make a quick decision, to listen to whoever knows the most and ask others to hold their tongues. But a team will become an amplification of its internal culture, for better or worse. Study after study shows that while psychological safety might be less efficient in the short run, it's more productive over time.

If motivation comes from giving individuals a greater sense of control, then psychological safety is the caveat we must remember when individuals come together in a group. Establishing control requires more than just seizing self-determination. Being a subversive works, unless you're leading a team.

When people come together in a group, sometimes we need to *give* control to others. That's ultimately what team norms are: individuals willingly giving a measure of control to their teammates. But that works only when people feel like they can trust one another. It only succeeds when we feel psychologically safe.

As a team leader, then, it's important to give people control. Some team leaders at Google make checkmarks next to people's names each time they speak, and won't end a meeting until those checks are all roughly equivalent. And as a team member, we share control by demonstrating that we are genuinely listening—by repeating what someone just said, by responding to their comments, by showing we care by reacting when someone seems upset or flustered, rather than acting as if nothing is wrong. When we defer to others' judgment, when we vocally treat others' concerns as our own, we give control to the group and psychological safety takes hold.

"The thing I love most is when I see a sketch performed and the actors are really killing it onstage, and the sketch's writers are high-fiving each other by the monitor, and whoever is waiting in the wings is laughing, and there's another team already figuring how to make the characters funnier next time," Lorne Michaels told me.

"When I see the entire team drawing some kind of inspiration from the same thing, I know everything is working," he said. "At that moment, the whole team is rooting for each other, and each person feels like the star."

FOCUS

Cognitive Tunneling, Air France Flight 447, and the Power of Mental Models

When they finally found the wreckage, it was clear that few of the victims had realized disaster was near even as it struck. There was no evidence of passengers' last-minute buckling of seatbelts or frenzied raising of food trays. Oxygen masks were firmly encased in ceiling panels. A submarine probing the wreckage at the bottom of the Atlantic Ocean found a whole row of seats upright in the sand, as if waiting to fly again.

It had taken almost two years to find the plane's data recorders and everyone hoped that once they were retrieved, the cause of the accident would become clear at last. Initially, however, the recorders offered few clues. None of the plane's computers had malfunctioned, according to the data. There was no indication of mechanical failure or electrical glitch. It wasn't until investigators listened to the cockpit voice recordings that they began to understand. This Airbus—one of the largest and most sophisticated aircraft ever built, a plane designed to be an error-proof model of automation—

was at the bottom of the ocean not because of a defect in machinery, but because of a failure of attention.

● ● ●

Twenty-three months earlier, on May 31, 2009, the night sky was clear as Air France Flight 447 pulled away from the gate in Rio de Janeiro with 228 people on board, bound for Paris. In the cabin were honeymooners and a former conductor for the Washington National Opera, a well-known arms control activist, and an eleven-year-old boy headed to boarding school. One of the plane's pilots had brought his wife to Rio so they could enjoy a three-day layover at the Copacabana Beach. Now she was in the back of the massive aircraft, while he and two colleagues were in the cockpit, flying them home.

As the plane began its ascent, there were a few radioed exchanges with air traffic control, the standard chatter that accompanies any takeoff. Four minutes after lifting from the runway, the pilot in the right seat—the junior position—activated the autopilot. For the next ten and a half hours, if all went according to plan, the plane would essentially fly itself.

Just two decades earlier, flying from Rio to Paris had been a much more taxing affair. Prior to the 1990s and advances in cockpit automation, pilots were responsible for calculating dozens of variables during a flight, including airspeed, fuel consumption, direction, and optimal cruising altitude, all while monitoring weather disturbances, discussions with air traffic control, and the plane's position in the sky. Such trips were so demanding that pilots often rotated responsibilities. They all knew the risks if vigilance waned. In 1987, a pilot in Detroit had become so overwhelmed during takeoff that he had forgotten to set the wing flaps. One hundred and fifty-four people died when the plane crashed after takeoff. Fifteen years before that, pilots flying near Miami had become fixated on a faulty landing gear light and had failed to notice that they were gradually descending. One hundred and one people were killed when the craft slammed into the Everglades. Before automated aviation systems were invented, it wasn't unheard of for more than a thousand people to die each year in airplane accidents, often because pilots' attention spans were stretched too thin, or due to other human errors.

The plane flying from Rio to Paris, however, had been designed to eliminate such mistakes by vastly reducing the number of decisions a pilot had to make. The Airbus A330 was so advanced that its computers could automatically intervene when problems arose, identify solutions, and then tell pilots, via on-screen instructions, where to direct their focus as they responded to computerized prompts. In optimal conditions, a human might fly for only about eight minutes per trip, during takeoff and landing. Planes like the A330 had fundamentally changed piloting from a proactive to a reactive profession. As a result, flying was easier. Accident rates went down, and airlines' productivity soared because more customers could travel with less crew. A transoceanic flight had once required as many as six pilots. By the time of Flight 447, thanks to automation, Air France needed only two people in the cockpit at any given time.

Four hours into the trip, midway between Brazil and Senegal, the plane crossed the equator. Most of the passengers would have been asleep. There were clouds from a tropical storm in the distance. The two men in the cockpit remarked on static electricity dancing across the windows, a phenomenon known as St. Elmo's fire. "I'm dimming the lighting a bit to see outside, eh?" said Pierre-Cedric Bonin, the pilot whose wife was in the passenger cabin. "Yes, yes," the captain replied. There was a third aviator in a small hold behind the cockpit, taking a nap. The captain summoned the third man to switch places, and then left the two junior pilots at the controls so he could sleep. The plane was flying smoothly on full autopilot at thirty-two thousand feet.

Twenty minutes later there was a small bump from turbulence. "It might be a good idea to tell the passengers to buckle up," Bonin informed a stewardess over the intercom. As the air surrounding the cockpit cooled, three metal cylinders jutting from the craft's body—the pitot tubes, which measure airspeed by detecting the force of air flowing into them—became clogged with ice crystals. For almost a hundred years, aviators have complained about, and safely accommodated, ice in pitot tubes. Most pilots know that if their airspeed measurement plunges unexpectedly, it's likely because of clogged pitot tubes. When the pitot tubes on Flight 447 froze over, the plane's computers lost airspeed information and the auto-flight system turned off, as it was programmed to do.

A warning alarm sounded.

"I have the controls," Bonin said calmly.

"Okay," his colleague replied.

At this point, if the aviators had done nothing at all, the plane would have continued flying safely and the pitot tubes would have eventually thawed. But Bonin, perhaps shaken out of a reverie by the alarm and wanting to offset the loss of the autopilot, pulled back a bit on the command stick, causing the plane's nose to nudge upward and the aircraft to gain altitude. Within one minute, it had ascended by three thousand feet.

With Flight 447's nose now pointed slightly upward, the plane's aerodynamics began to change. The atmosphere at that height was thin, and the ascent had disrupted the smooth flow of air over the plane's wings. The craft's "lift"—the basic force of physics that pulls airplanes into the sky because there is less pressure above a wing than below it—began deteriorating. In extreme conditions, this can cause an aerodynamic stall, a dangerous situation in which a plane starts falling, even as its engines strain with thrust and the nose points skyward. A stall is easy to overcome in its early stages. Simply lowering the nose so air begins flowing smoothly over the wings prevents a stall from emerging. But if a plane's nose remains upward, a stall will become worse and worse until the airplane drops like a stone in a well.

As Flight 447 rose through the thin atmosphere, a loud chime erupted in the cockpit and a recorded voice began warning, "Stall! Stall! Stall! Stall!," indicating that the plane's nose was pointed too high.

"What's this?" the copilot said.

"There's no good . . . uh . . . no good speed indication?" Bonin responded. The pitot tubes were still clogged with ice and so the display did not show any airspeed.

"Pay attention to your speed," the copilot said.

"Okay, okay, I'm descending," Bonin replied.

"It says we're going up," the copilot said, "so descend."

"Okay," said Bonin.

But Bonin didn't descend. If he had leveled the plane, the craft would have flown on safely. Instead, he continued pulling back on the stick slightly, pushing the airplane's nose further into the sky.

● ● ●

Automation has today penetrated nearly every aspect of our lives. Most of us now drive cars equipped with computers that automatically engage the brakes and reduce transmission power when we hit a patch of rain or ice, often so subtly we never notice the vehicle has anticipated our tendency to overcorrect. We work in offices where customers are routed to departments via computerized phone systems, emails are automatically sent when we're away from our desks, and bank accounts are instantaneously hedged against currency fluctuations. We communicate with smartphones that finish our words. Even without technology's help, all humans rely on cognitive automations, known as "heuristics," that allow us to multitask. That's why we can email the babysitter while chatting with our spouse and simultaneously watching the kids. Mental automation lets us choose, almost subconsciously, what to pay attention to and what to ignore.

Automations have made factories safer, offices more efficient, cars less accident-prone, and economies more stable. By one measure, there have been more gains in personal and professional productivity in the past fifty years than in the previous two centuries combined, much of it made possible by automation.

But as automation becomes more common, the risks that our attention spans will fail have risen. Studies from Yale, UCLA, Harvard, Berkeley, NASA, the National Institutes of Health, and elsewhere show errors are particularly likely when people are forced to toggle between automaticity and focus, and are unusually dangerous as automatic systems infiltrate airplanes, cars, and other environments where a misstep can be tragic. In the age of automation, knowing how to manage your focus is more critical than ever before.

Take, for instance, Bonin's mindset when he was forced to take control of Flight 447. It is unclear why he continued guiding the plane upward after agreeing with his copilot that they should descend. Maybe he hoped to climb above the storm clouds on the horizon. Perhaps it was an unintentional reaction to the sudden alarm. We will never know why he didn't return the controls to neutral once the stall warning sounded. There is significant evidence, however, that Bonin was in the grip of what's known as "cognitive tunneling"—a mental glitch that sometimes occurs when our brains are forced to transition abruptly from relaxed automation to panicked attention.

"You can think about your brain's attention span like a spotlight that can go wide and diffused, or tight and focused," said David Strayer, a cognitive psychologist at the University of Utah. Our attention span is guided by our intentions. We choose, in most situations, whether to focus the spotlight or let it be relaxed. But when we allow automated systems, such as computers or autopilots, to pay attention *for* us, our brains dim that spotlight and allow it to swing wherever it wants. This is, in part, an effort by our brains to conserve energy. The ability to relax in this manner gives us huge advantages: It helps us subconsciously control stress levels and makes it easier to brainstorm, it means we don't have to constantly monitor our environment, and it helps us get ready for big cognitive tasks. Our brains automatically seek out opportunities to disconnect and unwind.

"But then, *bam!,* some kind of emergency happens—or you get an unexpected email, or someone asks you an important question in a meeting—and suddenly the spotlight in your head has to ramp up all of a sudden and, at first, it doesn't know where to shine," said Strayer. "So the brain's instinct is to force it as bright as possible on the most obvious stimuli, whatever's right in front of you, even if that's not the best choice. That's when cognitive tunneling happens."

Cognitive tunneling can cause people to become overly focused on whatever is directly in front of their eyes or become preoccupied with immediate tasks. It's what keeps someone glued to their smartphone as the kids wail or pedestrians swerve around them on the sidewalk. It's what causes drivers to slam on their brakes when they see a red light ahead. We can learn techniques to get better at toggling between relaxation and concentration, but they require practice and a desire to remain engaged. However, once in a cognitive tunnel, we lose our ability to direct our focus. Instead, we latch on to the easiest and most obvious stimulus, often at the cost of common sense.

● ● ●

As the pitot tubes iced over and the alarms blared, Bonin entered a cognitive tunnel. His attention had been relaxed for the past four hours. Now, amid flashing lights and ringing bells, his attention searched for a focal point. The most obvious one was the video monitor right in front of his eyes.

The cockpit of an Airbus A330 is a minimalist masterpiece, an environment designed to be distraction free, with just a few screens alongside a modest number of gauges and controls. One of the most prominent screens, directly in each pilot's line of sight, is the primary flight display. There is a broad line running across the horizontal center of a screen that indicates the division between sky and land. Floating atop this line is the small icon of an aircraft. If a plane rolls to either side while flying, the icon goes off-kilter and pilots know their wings are no longer parallel to the ground.

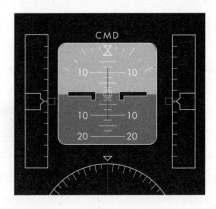

PRIMARY FLIGHT DISPLAY

When Bonin heard the alarm and looked at his instrument panel, he saw the primary flight display. The icon of the plane on that screen had rolled slightly to the right. Normally, this would not have been a concern. Planes roll in small increments throughout a trip and are easily righted. But now, with the autopilot disengaged and the sudden pressure to focus, the spotlight inside Bonin's head shined on that off-kilter icon. Bonin, data records indicate, became focused on getting the wings of that icon level with the middle of his screen. And then, perhaps because he was fixated on correcting the roll, he failed to notice that he was still pulling back on the control stick, lifting the plane's nose.

As Bonin pulled back on his stick, the front of the aircraft rose higher. Then, another instance of cognitive tunneling occurred—this time, inside the head of Bonin's copilot. The man in the left-hand seat was named David Robert, and he was officially the "monitoring pilot." His job was to keep watch over Bonin and intervene if the "flying pilot" became overwhelmed. In a worst-case scenario, Robert could take control of the craft. But now, with alarms sounding, Robert did what's most natural in such a situation: He became focused on the most obvious stimuli. There was a screen next to him spewing text as the plane's computer provided updates and instructions. Robert turned his eyes away from Bonin and began staring at the scrolling type, reading the messages aloud. "Stabilize," Robert said. "Go back down."

Focused on the screen as he was, Robert didn't see that Bonin was pulling back on his stick and didn't register that the flying pilot was raising the craft higher even as he agreed they needed to descend. There is no evidence that Robert looked at his gauges. Instead, he frantically scrolled through a series of messages automatically generated by the plane's computer. Even if those prompts had been helpful, nothing indicates that Bonin, locked on the airplane icon in front of him, heard anything his colleague said.

The plane rose through thirty-five thousand feet, drawing dangerously close to its maximum height. The nose of the airplane was now pitched at twelve degrees.

The copilot finally looked away from his screen. "We're climbing, according to this," he told Bonin, referring to the instrument panel. "Go back down!" he shouted.

"Okay," Bonin replied.

Bonin pushed his stick forward, forcing the plane's nose to dip slightly. As a result, the force of gravity against the pilots lessened by a third, giving them a brief sense of weightlessness. "Gently!" his colleague snapped. Then Bonin, perhaps overwhelmed by the combination of the alarms, the weightlessness, and his copilot's chastisement, jerked his hand backward, arresting the descent of the plane's nose. The craft remained at a six-degree upward pitch. Another loud warning chime came from the cockpit's speakers, and a few seconds later the aircraft began to shake, what's known as buf-feting, the result of rough air moving across the wings in the early stages of a serious aerodynamic stall.

"We're in, ahhh, yeah, we're in a climb, I think?" Bonin said.

For the next ten seconds, neither man spoke. The plane rose above its maximum recommended altitude of 37,500 feet. To stay aloft, Flight 447 *had* to descend. If Bonin would simply lower the nose, all would be fine.

Then, as the pilots focused on their screens, the ice crystals clog-ging the pitot tubes cleared and the plane's computer began receiv-ing accurate airspeed information once again. From this moment onward, all the craft's sensors functioned correctly throughout the flight. The computer began spitting out instructions, telling the pi-lots how to overcome the stall. Their instrument panels were show-ing them everything they needed to know to right the plane, but they had no idea where to look. Even as helpful information arrived, Bonin and Robert had no clue as to where to focus.

The stall warning blared again. A piercing, high-pitched noise called the "cricket," designed to be impossible for pilots to ignore, began to sound.

"Damn it!" the copilot yelled. He had already paged the captain. "Where is he? . . . Above all, try to touch the lateral controls as little as possible," he told Bonin.

"Okay," Bonin replied. "I'm in TO/GA, right?"

It is at this moment, investigators later concluded, that the lives of all 228 people on board Flight 447 were condemned. "TO/GA" is an acronym for "takeoff, go around," a setting that aviators use to abort a landing, or "go around" the runway. TO/GA pushes a plane's thrust to maximum while the pilot raises the nose. There is a se-quence of moves associated with TO/GA that all aviators practice, hundreds of times, in preparation for a certain kind of emergency. At low altitudes, TO/GA makes a lot of sense. The air is thick near the earth's surface, and so increasing thrust and raising the nose makes a plane go faster and higher, allowing a pilot to abort a land-ing safely.

But at thirty-eight thousand feet, the air is so thin that TO/GA doesn't work. A plane can't draw additional thrust at that height, and raising the nose simply increases the severity of a stall. At that altitude, the only correct choice is *lowering* the nose. In his star-tled panic, however, Bonin made a second mistake, a mental mis-step that is a cousin to cognitive tunneling: He sought to aim the spotlight in his head onto something familiar. Bonin fell back on a reaction he had practiced repeatedly, a sequence of moves he had learned to associate with emergencies. He fell into what psycholo-gists call "reactive thinking."

Reactive thinking is at the core of how we allocate our attention, and in many settings, it's a tremendous asset. Athletes, for example, practice certain moves again and again so that, during a game, they can think reactively and execute plays faster than their opponents can respond. Reactive thinking is how we build habits, and it's why to-do lists and calendar alerts are so helpful: Rather than needing to decide what to do next, we can take advantage of our reactive instincts and automatically proceed. Reactive thinking, in a sense, outsources the choices and control that, in other settings, create mo-tivation.

But the downside of reactive thinking is that habits and reactions can become so automatic they overpower our judgment. Once our motivation is outsourced, we simply react. One study conducted by Strayer, the psychologist, in 2009 looked at how drivers' behaviors changed when cars were equipped with features such as cruise con-trol and automatic braking systems that allowed people to pay less attention to road conditions.

"These technologies are supposed to make driving safer, and many times, they do," said Strayer. "But it also makes reactive think-ing easier, and so when the unexpected startles you, when the car skids or you have to brake suddenly, you'll react with practiced, habitual responses, like stomping on the pedal or twisting the wheel too far. Instead of thinking, you react, and if it's not the correct response, bad things happen."

● ● ●

Inside the cockpit, as the alarms sounded and the cricket chirped, the pilots were silent. Robert, the copilot, perhaps lost in his own thoughts, didn't reply to Bonin's question—"I'm in TO/GA, right?"—but instead tried once again to beckon the captain, who was still resting in the hold. If Bonin had paused to consider the basic facts—he was in thin air, a stall alarm was sounding, the plane couldn't safely go higher—he would have immediately realized he needed to lower the airplane's nose. Instead, he relied on behaviors he had practiced hundreds of times and pulled back on the stick. The plane's nose increased to a terrifying eighteen-degree pitch as Bonin pushed the throttle open. The plane moved higher, touched the top of an arc, and then started dropping, its nose still pointed upward and the engines at full thrust. The cockpit began shaking as the buffeting grew more pronounced. The plane was falling fast.

"What the hell is happening?" the copilot asked. "Do you understand what's happening, or not?"

"I don't have control of the plane anymore!" Bonin shouted. "I don't have control of the plane at all!"

In the cabin, passengers probably had little idea anything was wrong. There were no alarms they could hear. The buffeting likely felt like normal turbulence. Neither pilot ever made an announcement of any kind.

The captain finally entered the cockpit.

"What the hell are you doing?" he asked.

"I don't know what's happening," Robert said.

"We're losing control of the airplane!" Bonin shouted.

"We lost control of the airplane and we don't understand at all," Robert said. "We've tried everything."

Flight 447 was now sinking at a rate of ten thousand feet per minute. The captain, standing behind the pilots and perhaps overwhelmed by what he saw, uttered a curse word and then remained silent for forty-one seconds.

"I have a problem," Bonin said, the panic audible in his voice. "I have no more displays." This was not correct. The displays—the screens on his instrument panel—were providing accurate information and were clearly visible. But Bonin was too overwhelmed to focus.

"I have the impression we're going crazily fast," Bonin said. The plane, in fact, at this point was moving far too slowly. "What do you think?" Bonin asked as he reached for the lever that would raise the speed-brakes on the wing, slowing the plane even more.

"No!" shouted the copilot. "Above all, don't extend the brakes!"

"Okay," Bonin said.

"What should we do?" the copilot asked the captain. "What do you see?"

"I don't know," the captain said. "It's descending."

Over the next thirty-five seconds, as the pilots shouted questions, the plane dropped another nine thousand feet.

"Am I going down now?" Bonin asked. The instruments in front of him could have easily answered that question.

"You're going down down down," the copilot said.

"I've been at full back stick for a while," Bonin said.

"No, no!" the captain shouted. The plane was now less than ten thousand feet above the Atlantic Ocean. "Don't climb!"

"Give me the controls!" the copilot said. "The controls! To me!"

"Go ahead," Bonin says, finally releasing the stick. "You have the controls. We're still in TO/GA, right?"

As the copilot took over, the plane fell another six thousand feet closer to the sea.

"Watch out, you're pitching up there," the captain said.

"I'm pitching up?" the copilot replied.

"You're pitching up," the captain said.

"Well, we need to!" Bonin said. "We're at four thousand feet!"

By now, the only way the craft could pick up enough speed was to lower its nose into a dive and let more air flow over the wings. But with such a small distance between the plane and the ocean's surface, there was no room to maneuver. A ground proximity warning began blaring, "SINK RATE! PULL UP!" The cockpit was filled with constant noise.

"You're pitching up," the captain told the copilot.

"Let's go!" Bonin replied. "Pull up! Pull up! Pull up!"

The men stopped speaking for a moment.

"This can't be true," said Bonin. The ocean was visible through the cockpit's windows. If the pilots had craned their necks, they could have made out individual waves.

"But what's happening?" Bonin asked.

Two seconds later, the plane plunged into the sea.

II.

In the late 1980s, a group of psychologists at a consulting firm named Klein Associates began exploring why some people seem to stay calm and focused amid chaotic environments while others become overwhelmed. Klein Associates' business was helping companies analyze how they make decisions. A variety of clients wanted to know why some employees made such good choices amid stress and time pressures, while other workers became distracted. More important, they wanted to know if they could train people to get better at paying attention to the right things.

The Klein Associates team began by interviewing professionals who worked in extreme settings, such as firefighters, military commanders, and emergency rescue personnel. Many of those conversations, however, proved frustrating. Firefighters could look at a burning staircase and sense if it would hold their weight, they knew which parts of a building needed constant attention and how to stay attuned to warning signs, but they struggled to explain how they did it. Soldiers could tell you which parts of a battlefield were more likely to be harboring enemies and where to focus for signs of ambush. But when asked to explain their decisions, they chalked it up to intuition.

So the team moved on to other settings. One researcher, Beth Crandall, visited neonatal intensive care units, or NICUs, around Dayton, near where she lived. A NICU, like all critical care settings, is a mix of chaos and banality set against a backdrop of constantly beeping machines and chiming warnings. Many of the babies inside a NICU are on their way to full health; they might have arrived prematurely or suffered minor injuries during birth, but they are not seriously ill. Others, though, are unwell and need constant monitoring. What makes things particularly hard for NICU nurses, however, is that it is not always clear which babies are sick and which are healthy. Seemingly okay preemies can become unwell quickly; sick infants can recover unexpectedly. So nurses are constantly making choices about where to focus their attention: the squalling baby or the quiet one? The new lab results or the worried parents who say something seems wrong? What's more, these choices occur amid a constant stream of data from machines—heart monitors and automatic thermometers, blood pressure systems and pulse oximeters—that are ready to sound alarms the moment anything changes. Such innovations have made patients safer and have remarkably improved NICUs' productivity, because fewer nurses are now needed to oversee greater numbers of children. But they have also made NICUs more complex. Crandall wanted to understand how nurses made decisions about which babies needed their attention, and why some of them were better at focusing on what mattered most.

Crandall interviewed nurses who were calm in the face of emergencies and others who seemed on the brink of collapse. Most interesting were the handful of nurses who seemed particularly gifted at noticing when a baby was in trouble. They could predict an infant's decline or recovery based on small warning signs that almost everyone else overlooked. Often, the clues these nurses relied upon to spot problems were so subtle that they themselves had trouble later recalling what had prompted them to act. "It was like they could see things no one else did," Crandall told me. "They seemed to think differently."

One of Crandall's first interviews was with a talented nurse named Darlene, who described a shift from a few years earlier. Darlene had been walking past an incubator when she happened to glance at the baby inside. All of the machines hooked up to the child showed that her vitals were within normal ranges. There was another RN keeping watch over the baby, and she was observing the infant attentively, unconcerned by what she saw. But to Darlene, something seemed wrong. The baby's skin was slightly mottled instead of uniformly pink. The child's belly seemed a bit distended. Blood had recently been drawn from a pinprick in her heel and the Band-Aid showed a blot of crimson, rather than a small dot.

None of that was particularly unusual or troubling. The nurse tending to the child said she was eating and sleeping well. Her heartbeat was strong. But something about all those small things occurring together caught Darlene's attention. She opened the incubator and examined the infant. The newborn was conscious and awake. She grimaced slightly at Darlene's touch but didn't cry. There was nothing specific that she could point to, but this baby simply didn't look like Darlene expected her to.

Darlene found the attending physician and said they needed to start the child on intravenous antibiotics. All they had to go on was Darlene's intuition, but the doctor, deferring to her judgment, ordered the medication and a series of tests. When the labs came back, they showed that the baby was in the early stages of sepsis, a potentially fatal whole-body inflammation caused by a severe infection. The condition was moving so fast that, had they waited any longer, the newborn would have likely died. Instead, she recovered fully.

"It fascinated me that Darlene and this other nurse had seen the same warning signs, they had all the same information, but only Darlene detected the problem," Crandall said. "To the other nurse, the mottled skin and the bloody Band-Aid were data points, nothing big enough to trigger an alarm. But Darlene put everything together. She saw a whole picture." When Crandall asked Darlene to explain how she knew the baby was sick, Darlene said it was a hunch. As Crandall asked more questions, however, another explanation emerged. Darlene explained that she carried around a picture in her mind of what a healthy baby *ought* to look like—and the infant in the crib, when she glanced at her, hadn't matched that image. So the spotlight inside Darlene's head went to the child's skin, the blot of blood on her heel, and the distended belly. It focused on those unexpected details and triggered Darlene's sense of alarm. The other nurse, in contrast, didn't have a strong picture in her head of what she expected to see, and so her spotlight focused on the most obvious details: The baby was eating. Her heartbeat was strong. She wasn't crying. The other nurse was distracted by the information that was easiest to grasp.

People like Darlene who are particularly good at managing their attention tend to share certain characteristics. One is a propensity to create pictures in their minds of what they expect to see. These people tell themselves stories about what's going on as it occurs. They narrate their own experiences within their heads. They are more likely to answer questions with anecdotes rather than simple responses. They say when they daydream, they're often imagining future conversations. They visualize their days with more specificity than the rest of us do.

Psychologists have a phrase for this kind of habitual forecasting:

"creating mental models." Understanding how people build mental models has become one of the most important topics in cognitive psychology. All people rely on mental models to some degree. We all tell ourselves stories about how the world works, whether we realize we're doing it or not.

But some of us build more robust models than others. We envision the conversations we're going to have with more specificity, and imagine what we are going to do later that day in greater detail. As a result, we're better at choosing where to focus and what to ignore. The secret of people like Darlene is that they are in the habit of telling themselves stories all the time. They engage in constant forecasting. They daydream about the future and then, when life clashes with their imagination, their attention gets snagged. That helps explain why Darlene noticed the sick baby. She was in the habit of imagining what the babies in her unit ought to look like. Then, when she glanced over and the bloody Band-Aid, distended belly, and mottled skin didn't match the image in her mind, the spotlight in her head swung toward the child's bassinet.

Cognitive tunneling and reactive thinking occur when our mental spotlights go from dim to bright in a split second. But if we are constantly telling ourselves stories and creating mental pictures, that beam never fully powers down. It's always jumping around inside our heads. And, as a result, when it has to flare to life in the real world, we're not blinded by its glare.

● ● ●

When the Air France Flight 447 investigators began parsing cockpit audio recordings, they found compelling evidence that none of the pilots had strong mental models during their flight.

"What's this?" the copilot asked when the first stall warning sounded.

"There's no good speed indication? . . . We're in . . . we're in a climb?" Bonin responded.

The pilots kept asking each other questions as the plane's crisis deepened because they didn't have mental models to help them process new information as it arrived. The more they learned, the more confused they became. This explains why Bonin was so prone to cognitive tunneling. He hadn't been telling himself a story as the plane flew along, and so when the unexpected occurred, he wasn't sure which details to pay attention to. "I have the impression we're going crazily fast," he said as the plane began to slow and fall. "What do you think?"

And when Bonin did finally latch on to a mental model—"I'm in TO/GA, right?"—he didn't look for any facts that challenged that model. "I'm climbing, okay, so we're going down," he said two minutes before the plane crashed, seemingly oblivious to the contradiction of his words. "Okay, we're in TO/GA," he added. "How come we're continuing to go right down?"

"This can't be true," he said seconds before the plane hit the water. Then there are his last words, which make all the sense in the world once you realize Bonin was still grasping for useful mental models as the plane hurtled toward the waves:

"But what's happening?"

This problem isn't unique to the aviators of Flight 447, of course. It happens all the time, within offices and on freeways, as we're working on our smartphones and multitasking from the couch. "This mess of a situation is one hundred percent our own fault," said Stephen Casner, a research psychologist at NASA who has studied dozens of accidents like Air France Flight 447. "We started with a creative, flexible, problem-solving human and a mostly dumb computer that's good at rote, repetitive tasks like monitoring. So we let the dumb computer fly and the novel-writing, scientific-theorizing, jet-flying humans sit in front of the computer like potted plants watching for blinking lights. It's always been difficult to learn how to focus. It's even harder now."

●●●

A decade after Beth Crandall interviewed the NICU nurses, two economists and a sociologist from MIT decided to study how, exactly, the most productive people build mental models. To do that, they convinced a midsized recruiting firm to give them access to their profit-and-loss data, employees' appointment calendars, and the 125,000 email messages the firm's executives had sent over the previous ten months.

The first thing the researchers noticed, as they began crawling through all that data, was that the firm's most productive workers, its superstars, shared a number of traits. The first was they tended to work on only five projects at once—a healthy load, but not extraordinary. There were other employees who handled ten or twelve projects at a time. But those employees had a lower profit rate than the superstars, who were more careful about how they invested their time.

The economists figured the superstars were pickier because they were seeking out assignments that were similar to previous work they had done. Conventional wisdom holds that productivity rises when people do the same kind of tasks over and over. Repetition makes us faster and more efficient because we don't have to learn fresh skills with each new assignment. But as the economists looked more closely, they found the opposite: The superstars weren't choosing tasks that leveraged existing skills. Instead, they were signing up for projects that required them to seek out new colleagues and demanded new abilities. That's why the superstars worked on only five projects at a time: Meeting new people and learning new skills takes a lot of additional hours.

Something else the superstars had in common is they were disproportionately drawn to assignments that were in their early stages. This was surprising, because joining a project in its infancy is risky. New ideas often fail, no matter how smart or well executed. The safest bet is signing on to a project that is well under way.

However, the beginning of a project is also more information rich. By joining fledgling initiatives, the superstars were cc'd on emails they wouldn't have otherwise seen. They learned which junior executives were smart and picked up new ideas from their younger colleagues. They were exposed to emerging markets and the lessons of the digital economy earlier than other executives. What's more, the superstars could later claim ownership of an innovation simply by being in the room when it was born, rather than fighting paternity battles once it was deemed a success.

Finally, the superstars also shared a particular behavior, almost an intellectual and conversational tic: They loved to generate theories—lots and lots of theories, about all kinds of topics, such as why certain accounts were succeeding or failing, or why some clients were happy or disgruntled, or how different management styles influenced various employees. They were somewhat obsessive, in fact, about trying to explain the world to themselves and their colleagues as they went about their days.

The superstars were constantly telling stories about what they had seen and heard. They were, in other words, much more prone to generate mental models. They were more likely to throw out ideas during meetings, or ask colleagues to help them imagine how future conversations might unfold, or envision how a pitch should go. They came up with concepts for new products and practiced how they would sell them. They told anecdotes about past conversations and dreamed up far-fetched expansion plans. They were building mental models at a near constant rate.

"A lot of these people will come up with explanation after explanation about what they just saw," said Marshall Van Alstyne, one of the MIT researchers. "They'll reconstruct a conversation right in front of you, analyzing it piece by piece. And then they'll ask you to challenge them on their take. They're constantly trying to figure out how information fits together."

The MIT researchers eventually calculated that getting cc'd on those early information-rich emails and hashing out those mental models earned the superstars an extra $10,000 a year, on average, in bonuses. The superstars took on only five projects at once—but they outperformed their colleagues because they had more productive methods of thinking.

Researchers have found similar results in dozens of other studies. People who know how to manage their attention and who habitually build robust mental models tend to earn more money and get better grades. Moreover, experiments show that anyone can learn to habitually construct mental models. By developing a habit of telling ourselves stories about what's going on around us, we learn to sharpen where our attention goes. These storytelling moments can be as small as trying to envision a coming meeting while driving to work—forcing yourself to imagine how the meeting will start, what points you will raise if the boss asks for comments, what objections your coworkers are likely to bring up—or they can be as big as a nurse telling herself stories about what infants ought to look like as she walks through a NICU.

If you want to make yourself more sensitive to the small details in your work, cultivate a habit of imagining, as specifically as possible, what you expect to see and do when you get to your desk. Then you'll be prone to notice the tiny ways in which real life deviates from the narrative inside your head. If you want to become better at listening to your children, tell yourself stories about what they said to you at dinnertime last night. Narrate your life, as you are living it, and you'll encode those experiences deeper in your brain. If you need to improve your focus and learn to avoid distractions, take a moment to visualize, with as much detail as possible, what you are about to do. It is easier to know what's ahead when there's a well-rounded script inside your head.

Companies say such tactics are important in all kinds of settings, including if you're applying for a job or deciding whom to hire. The candidates who tell stories are the ones every firm wants. "We look for people who describe their experiences as some kind of a narrative," Andy Billings, a vice president at the video game giant Electronic Arts, told me. "It's a tip-off that someone has an instinct for connecting the dots and understanding how the world works at a deeper level. That's who everyone tries to get."

III.

One year after Air France Flight 447 disappeared into the ocean, another Airbus—this one part of Qantas Airways—taxied onto a runway in Singapore, requested permission to begin the eight-hour flight to Sydney, and lifted into the bright morning sky.

The Qantas plane flying that day had the same auto-flight systems as the Air France airplane that had crashed into the sea. But the pilots were very different. Even before Captain Richard Champion de Crespigny stepped on board Qantas Flight 32, he was drilling his crew in the mental models he expected them to use.

"I want us to envision the first thing we'll do if there's a problem," he told his copilots as they rode in a van from the Fairmont hotel to Singapore Changi Airport. "Imagine there's an engine failure. Where's the first place you'll look?" The pilots took turns describing where they would turn their eyes. De Crespigny conducted this same conversation prior to every flight. His copilots knew to expect it. He quizzed them on what screens they would stare at during an emergency, where their hands would go if an alarm sounded, whether they would turn their heads to the left or stare straight ahead. "The reality of a modern aircraft is that it's a quarter million sensors and computers that sometimes can't tell the difference between garbage and good sense," de Crespigny later told me. He's a brusque Australian, a cross between Crocodile Dundee and General Patton. "That's why we have human pilots. It's our job to think about what *might* happen, instead of what is."

After the crew's visualization session, de Crespigny laid down some rules. "Everyone has a responsibility to tell me if you disagree with my decisions or think I'm missing anything."

"Mark," he said, gesturing to a copilot, "if you see everyone looking down, I want you to look up. If we're all looking up, you look down. We'll all probably make at least one mistake this flight. You're each responsible for catching them."

Four hundred and forty passengers were preparing to board the plane when the pilots entered the cockpit. De Crespigny, like all Qantas aviators, was required to undergo a yearly review of his flying skills, and so, on that day, there were two extra pilots in the cockpit, observers drawn from the airline's most experienced ranks. The review wasn't perfunctory. If de Crespigny stumbled, it could trigger his early retirement.

As the pilots took their seats, one of the observers sat near the center of the cockpit, where standard operating procedure usually positioned the second officer. De Crespigny frowned. He had expected the observer to sit off to the side, out of the way. He had a picture in his mind of how his cockpit ought to be arranged.

De Crespigny faced the evaluator. "Where do you intend to sit?" he asked.

"In this seat between you and Matt," the observer said.

"I've got a problem with that," de Crespigny said. "You're inhibiting my crew."

The cockpit went silent. This kind of confrontation was not supposed to happen between a captain and the observers.

"Rich, I can't see you if I sit in Mark's seat," the observer said. "How can I check you?"

"That's your problem," de Crespigny replied. "I want my crew together and I want Mark in your seat."

"Richard, you're being unreasonable," the second observer said.

"I have a flight to command and I want my crew operating properly," said de Crespigny.

"Look, Richard," replied the evaluator, "if it helps, I promise I'll be the second officer if I have to be."

De Crespigny paused. He wanted to show his crew they could question his decisions. He wanted them to know he was paying close attention to what they had to say and was sensitive to what they thought. Just as teams at Google and *Saturday Night Live* need to be able to critique one another without fear of punishment, de Crespigny wanted his crew to see that he encouraged them to disagree.

"Fantastic," de Crespigny said to the evaluator. ("Once he said he would be the second officer, it fit into the plan I had in my mind," de Crespigny later told me.) Inside the cockpit, de Crespigny turned back to the controls and began moving Qantas Flight 32 away from the gate.

The plane sped down the runway and lifted into the air. At 2,000 feet, de Crespigny activated the plane's autopilot. The sky was cloudless, the conditions perfect.

At 7,400 feet, as de Crespigny was about to order the first officer to switch off the cabin's seatbelt sign, he heard a boom. It was probably just a surge of high-pressure air moving through the engine, he thought. Then there was another, even louder crash, followed by what sounded like thousands of marbles being thrown against the hull.

A red alarm flashed on de Crespigny's instrument panel and a siren blared in the cockpit. Investigators would later determine that an oil fire inside one of the left jets had caused a massive turbine disk to detach from the drive shaft, shear into three pieces, and shoot outward, shattering the engine. Two of the larger fragments from that explosion punched holes in the left wing, one of them large enough for a man to fit through. Hundreds of smaller shards, exploding like a cluster bomb, cut through electrical wires, fuel hoses, a fuel tank, and hydraulic pumps. The underside of the wing looked as though it had been machine-gunned.

Long strips of metal were bending off the left wing and whipping in the air. The plane began to shake. De Crespigny reached over to decrease the aircraft's speed, the standard reaction for an emergency of this kind, but when he pushed a button, the auto-thrust didn't respond. Alarms started popping up on his computer display. Engine two was on fire. Engine three was damaged. There was no data at all for engines one and four. The fuel pumps were failing. The hydraulics, pneumatics, and electrical systems were almost inoperative. Fuel was leaking from the left wing in a wide fan. The damage would later be described as one of the worst midair mechanical disasters in modern aviation.

De Crespigny radioed Singapore air traffic control. "QF32, engine two appears failed," he said. "Heading 150, maintaining 7,400 feet, we'll keep you informed and will get back to you in five minutes."

Less than ten seconds had passed since the first boom. De Crespigny cut power to the left wing and began anti-fire protocols. The plane stopped vibrating for a moment. Inside the cockpit, alarms were blaring. The pilots were quiet.

In the cabin, panicked passengers rushed to their windows and pointed at the screens embedded in their seats, which, unfortunately, were broadcasting the view of the damaged wing from a camera mounted in the tail.

The men in the cockpit began responding to prompts from the plane's computers, speaking to one another in short, efficient sentences. De Crespigny looked at his display and saw that twenty-one of the plane's twenty-two major systems were damaged or completely disabled. The functioning engines were rapidly deteriorating and the left wing was losing the hydraulics that made steering possible. Within minutes, the plane had become capable of only the smallest changes in thrust and the tiniest navigational adjustments. No one was certain how long it would stay in the air.

One of the copilots looked up from his controls. "I think we should turn back," he said. Turning the airplane around in order to head back to the airport was risky. But at their current heading, they were getting farther away from the runway with each second.

De Crespigny told the control tower they would return. He began turning the plane in a long, slow arc. "Request climb to ten thousand feet," de Crespigny radioed to air traffic control.

"No!" his copilots shouted.

They quickly explained their concerns: Climbing higher might strain the engines. The change in altitude could cause fuel to leak faster. They wanted to stay low and keep the plane flat.

De Crespigny had flown more than fifteen thousand hours as a pilot and had practiced disaster scenarios like this in dozens of simulators. He had envisioned moments like this hundreds of times. He had a picture in his mind of how to react, and it involved getting higher so he would have more options. Every instinct told him to gain altitude. But each mental model has gaps. It was his crew's job to find them.

"Qantas 32," de Crespigny radioed. "Disregard the climb to 10,000 feet. We will maintain 7,400 feet."

● ● ●

For the next twenty minutes, the men in the cockpit dealt with an increasing number of alarms and emergencies. The plane's computer displayed step-by-step solutions to each problem, but as the issues cascaded, the instructions became so overwhelming that no one was certain how to prioritize or where to focus. De Crespigny felt himself getting drawn into a cognitive tunnel. One computer checklist told the pilots to transfer fuel between the wings in order to balance the plane's weight. "Stop!" de Crespigny shouted as a copilot reached to comply with the screen's command. "Should we be transferring fuel out of the good right wing into the leaking left wing?" A decade earlier, a flight in Toronto had nearly crashed after the crew had inadvertently dumped their fuel by transferring it into a leaky engine. The pilots agreed to ignore the order.

De Crespigny slumped in his chair. He was trying to visualize the damage, trying to keep track of his dwindling options, trying to construct a mental picture of the plane as he learned more and more about what was wrong. Throughout this crisis, de Crespigny and the other pilots had been building mental models of the Airbus inside their heads. Everywhere they looked, however, they saw a new alarm, another system failing, more blinking lights. De Crespigny took a breath, removed his hand from the controls, and placed them in his lap.

"Let's keep this simple," he said to his copilots. "We can't transfer fuel, we can't jettison it. The trim tank fuel is stuck in the tail and the transfer tanks are useless.

"So forget the pumps, forget the other eight tanks, forget the total fuel quantity gauge. We need to stop focusing on what's wrong and start paying attention to what's still working."

On cue, one of the copilots began ticking off things that were still operational: Two of eight hydraulic pumps still functioned. The left wing had no electricity, but the right wing had some power. The wheels were intact and the copilots believed de Crespigny could pump the brakes at least once before they failed.

The first airplane de Crespigny had ever flown was a Cessna, one of the single-engine, nearly noncomputerized planes that hobbyists loved. A Cessna is a toy compared to an Airbus, of course, but every plane, at its core, has the same components: a fuel system, flight controls, brakes, landing gear. *What if,* de Crespigny thought to himself, *I imagine this plane as a Cessna? What would I do then?*

"That moment is really the turning point," Barbara Burian, a research psychologist at NASA who has studied Qantas Flight 32, told me. "When de Crespigny decided to take control of the mental model he was applying to the situation, rather than react to the computer, it shifted his mindset. Now, he's deciding where to direct his focus instead of relying on instructions.

"Most of the time, when information overload occurs, we're not aware it's happening—and that's why it's so dangerous," Burian said. "So really good pilots push themselves to do a lot of 'what if' exercises before an event, running through scenarios in their heads. That way, when an emergency happens, they have models they can use."

This shift in mindset—*What if I imagine this plane as a Cessna?*—is what never occurred, tragically, inside the cockpit of Air France Flight 447. The French pilots never reached for a new mental model to explain what was going on. But when the mental model of the Airbus inside de Crespigny's head started coming apart under the weight of all the new emergencies, he decided to replace it with something new. He began imagining the plane as a Cessna, which allowed him to figure out where he should turn his attention and what he could ignore.

De Crespigny asked one of his copilots to calculate how much runway they would need. Inside his head, de Crespigny was envisioning the landing of an oversized Cessna. "Picturing it that way helped me simplify things," he told me. "I had a picture in my head that contained the basics, and that's all I needed to land the plane."

If de Crespigny hit everything just right, the copilot said, the plane would require 3,900 meters of asphalt. The longest runway at Singapore Changi was 4,000 meters. If they overshot, the craft would buckle as its wheels hit the grassy fields and sand dunes.

"Let's do this," de Crespigny said.

The plane began descending toward Singapore Changi airport. At two thousand feet, de Crespigny looked up from his panel and saw the runway. At one thousand feet, an alarm inside the cockpit began screaming "SPEED! SPEED! SPEED!" The plane was at risk of stalling. De Crespigny's eyes flicked between the runway and his speed indicators. He could see the Cessna's wings in his mind. He delicately nudged the throttle, increasing the speed slightly, and the alarm stopped. He brought the nose up a touch because that's what the picture in his mind told him to do.

"Confirm the fire services on standby," a copilot radioed the control tower.

"Affirm, we have the emergency services on standby," a voice replied.

The plane was descending at fourteen feet per second. The maximum certified speed the undercarriage could absorb was only twelve feet per second. But there were no other options now.

"FIFTY," a computerized voice said. "FORTY." De Crespigny pulled back slightly on his stick. "THIRTY . . . TWENTY." A metallic voice erupted: "STALL! STALL! STALL!" The Cessna in de Crespigny's mind was still sailing toward the runway, ready to land as he had hundreds of times before. It wasn't stalling. He ignored the alarm. The rear wheels of the Airbus touched the ground and de Crespigny pushed his stick forward, forcing the front wheels onto the tarmac. The brakes would work only once, so de Crespigny pushed the pedal as far as it would go and held it down. The first thousand meters of the runway blurred past. At the two-thousand-meter mark, de Crespigny thought they might be slowing. The end of the runway was rushing toward them through the windshield, grass and sand dunes growing bigger the closer they got. As the plane neared the end of the runway, the metal began to groan. The wheels left long skid marks on the asphalt. Then the plane slowed, shuddered, and came to a stop with one hundred meters to spare.

Investigators would later deem Qantas Flight 32 the most damaged Airbus A380 ever to land safely. Multiple pilots would try to re-create de Crespigny's recovery in simulators and would fail every time.

When Qantas Flight 32 finally came to a rest, the lead flight attendant activated the plane's announcement system.

"Ladies and gentlemen," he said, "welcome to Singapore. The local time is five minutes to midday on Thursday 4 November, and I think you'll agree that was one of the nicest landings we have experienced for a while." De Crespigny returned home a hero. Today, Qantas Flight 32 is taught in flight schools and psychology classrooms as a case study of how to maintain focus during an emergency. It is cited as one of the prime examples of how mental models can put even the most dire situations within our control.

Mental models help us by providing a scaffold for the torrent of information that constantly surrounds us. Models help us choose where to direct our attention, so we can make decisions, rather than just react. The Air France pilots didn't have strong mental models, and so when tragedy struck, they didn't know where to focus. De Crespigny and his copilots, in contrast, were telling themselves stories—and testing and revising them—even before they stepped onto the plane, and so they were prepared when disaster occurred.

We may not recognize how situations within our own lives are similar to what happens within an airplane cockpit. But think, for a moment, about the pressures you face each day. If you are in a meeting and the CEO suddenly asks you for an opinion, your mind is likely to snap from passive listening to active involvement—and if you're not careful, a cognitive tunnel might prompt you to say something you regret. If you are juggling multiple conversations and tasks at once and an important email arrives, reactive thinking can cause you to type a reply before you've really thought out what you want to say.

So what's the solution? If you want to do a better job of paying attention to what really matters, of not getting overwhelmed and

distracted by the constant flow of emails and conversations and in-
terruptions that are part of every day, of knowing where to focus and
what to ignore, get into the habit of telling yourself stories. Narrate
your life as it's occurring, and then when your boss suddenly asks
a question or an urgent note arrives and you have only minutes to
reply, the spotlight inside your head will be ready to shine the right
way.

To become genuinely productive, we must take control of our attention; we must build mental models that put us firmly in charge.
When you're driving to work, force yourself to envision your day.
While you're sitting in a meeting or at lunch, describe to yourself
what you're seeing and what it means. Find other people to hear
your theories and challenge them. Get in a pattern of forcing yourself to anticipate what's next. If you are a parent, anticipate what
your children will say at the dinner table. Then you'll notice what
goes unmentioned or if there's a stray comment that you should see
as a warning sign.

"You can't delegate thinking," de Crespigny told me. "Computers
fail, checklists fail, everything can fail. But people can't. We have to
make decisions, and that includes deciding what deserves our attention. The key is forcing yourself to think. As long as you're thinking,
you're halfway home."

GOAL SETTING

Smart Goals, Stretch Goals, and the Yom Kippur War

In October 1972, one of Israel's brightest generals, the forty-four-year-old Eli Zeira, was promoted to oversee the Directorate of Military Intelligence, the agency responsible for warning the country's leaders if its enemies were about to attack.

Zeira's appointment came half a decade after the 1967 Six-Day War, in which Israel, in a stunning preemptive strike, had captured the Sinai Peninsula, the Golan Heights, and other territory from Egypt, Syria, and Jordan. That war had demonstrated Israel's military superiority, more than doubled the amount of territory the country controlled, and had humiliated the nation's enemies. But it also instilled a deep anxiety among Israeli citizens that the country's antagonists would eventually seek revenge.

Those fears were legitimate. Since the Six-Day War had ended, generals in Egypt and Syria had repeatedly threatened to reclaim their lost territory, and Arab leaders, in fiery speeches, had vowed to push the Jewish state into the sea. As Israel's enemies became increasingly bellicose, the nation's lawmakers sought to calm public worries by asking the military to provide regular forecasts on the likelihood of attack.

However, the assessments provided by the Directorate of Military Intelligence were often contradictory and inconclusive, a mishmash of opinions predicting various levels of risk. Analysts sent conflicting memos and flip-flopped week to week. Some weeks, lawmakers were warned to be on alert, and then nothing would happen. Policy makers were called to meetings and told that a risk might be materializing, but no one could say for sure. Army units were ordered to ready their defenses, and then those orders would be countermanded with no explanation why.

As a result, Israeli politicians and the public became increasingly frustrated. Army reservists constituted 80 percent of the Israeli Defense Forces' ground troops. There was a constant nervousness that hundreds of thousands of citizens would be required, at a moment's notice, to abandon their families and rush to the borders. People wanted to know if the risk of another war was real and, if so, how much forewarning they would get.

Eli Zeira was appointed to head the Directorate of Military Intelligence, in part, to address those uncertainties. He was a former paratrooper known for his sophistication and political savvy. He had risen quickly through Israel's military establishment, even spending a few years as an assistant to Moshe Dayan, the hero of the Six-Day War. When Zeira took over the Directorate, he told the Israeli parliament that his job was simple: to provide decision makers with an "estimate as clear and as sharp as possible." His chief goal, he said, was to make sure alarms were raised only when the risks of war were real.

His method for achieving this clarity was ordering his military analysts to use a strict formula in assessing Arab intentions. He had helped develop these criteria, which became known among intelligence officials as "the concept." Zeira argued that during the Six-

Day War, Israel's superior airpower, arsenal of long-range missiles, and battlefield dominance had so thoroughly embarrassed their enemies that no country would attack again unless they had an air force powerful enough to protect ground troops from Israeli jets, and Scud missiles capable of hitting Tel Aviv. Without those two conditions being met, Zeira said, the threats of Arab leaders were nothing more than words.

Six months after Zeira assumed his post, the nation had an opportunity to test his concept. In the spring of 1973, large numbers of Egyptian troops began amassing along the Suez Canal, which was the border between Egypt and the Israeli-controlled Sinai Peninsula. Israel's spies warned that Egypt planned to invade in mid-May.

On April 18, Israel's prime minister, Golda Meir, gathered her top advisers in a closed-door meeting. The military chief of staff and the head of the Mossad both said an Egyptian attack was a real possibility and the nation needed to prepare.

Meir turned to Zeira for his assessment. He disagreed with his colleagues, he said. Egypt still didn't have a powerful air force and possessed no missiles capable of reaching Tel Aviv. Egypt's leaders were merely rattling sabers to impress their countrymen. The odds of an invasion, he determined, were "very low."

Meir ultimately sided with her chief of staff and

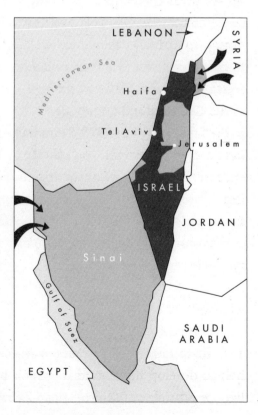

the Mossad. She ordered the military to make defensive preparations and, over the next month, the army readied itself for war.
Soldiers built walls, outposts, and batteries along the hundred-mile-
long bank of the Suez Canal. In the Golan Heights, which bordered
Syria, platoons launched practice shells and tanks rehearsed battle
formations. Millions of dollars were spent and thousands of soldiers
were prevented from taking leave. But the attack never materialized.
Meir's government, chagrined at their overreaction, soon reversed
their public declarations. In July of that year, Moshe Dayan, then
Israel's defense minister, told *Time* magazine that it was unlikely
a war would occur within the next decade. Zeira emerged from the
affair, in the words of the historian Abraham Rabinovich, "with his
reputation, and his self-confidence, greatly enhanced.

"With alarm bells going off all around him and the nation's fate at
stake, he had coolly maintained throughout the crisis that the probability of war was not only low, but 'very low,'" Rabinovich wrote.
"It was [his] task, he would say, to keep the national blood pressure
down and not sound alarms unnecessarily. Otherwise, the reserves
would be mobilized every couple of months with devastating effect
on the economy and on morale."

By the summer of 1973, Zeira had established himself as one of
Israel's most influential leaders. He had assumed his new job with
the goal of reducing needless anxiety, and had demonstrated that a
disciplined approach could prevent wasteful second-guessing. The
nation had wanted relief from the constant worries of an impending
attack, and Zeira had provided it. His ascent to even more powerful
positions seemed preordained.

II.

Imagine you have been asked to complete a questionnaire. Your assignment is to rate how strongly you agree or disagree with forty-
two statements, including:

I believe orderliness and organization are among the most
   important characteristics.
I find that establishing a consistent routine enables me to
   enjoy life more.
I like to have friends who are unpredictable.
I prefer interacting with people whose opinions are very dif-
   ferent from my own.
My personal space is usually messy and disorganized.
It's annoying to listen to someone who cannot seem to make
   up his or her mind.

A team of researchers at the University of Maryland first published this test in 1994, and since then it has become a staple of
personality exams. At first glance, the questions seem designed to
measure someone's preference for personal organization and their
comfort with alternate viewpoints. And, in fact, researchers have
found that this exam helps identify people who are more decisive
and self-assured, and that those traits are correlated with general
success in life. Determined and focused people tend to work harder
and get tasks done more promptly. They stay married longer and
have deeper networks of friends. They often have higher-paying
jobs.

But this questionnaire is not intended to test personal organization. Rather, it's designed to measure a personality trait known as
"the need for cognitive closure," which psychologists define as "the
desire for a confident judgment on an issue, any confident judgment,
as compared to confusion and ambiguity." Most people respond to
this exam—which is called "the need for closure scale"—by demonstrating a preference for a mix of order and chaos in their lives. They
say they prize orderliness but admit to having messy desks. They
say they are annoyed by indecision but also have unreliable friends.
However, some people—about 20 percent of test takers, and many
of the most accomplished people who have completed the exam—

show a higher-than-average preference for personal organization, decisiveness, and predictability. They tend to disdain flighty friends and ambiguous situations. These people have a high emotional need for cognitive closure.

The need for cognitive closure, in many settings, can be a great strength. People who have a strong urge for closure are more likely to be self-disciplined and seen as leaders by their peers. An instinct to make a judgment and then stick with it forestalls needless second-guessing and prolonged debate. The best chess players typically display a high need for closure, which helps them focus on a specific problem during stressful moments rather than obsessing over past mistakes. All of us crave closure to some degree, and that's good, because a basic level of personal organization is a prerequisite for success. What's more, making a decision and moving on to the next question *feels* productive. It feels like progress.

But there are risks associated with a high need for closure. When people begin craving the emotional satisfaction that comes from making a decision—when they require a sensation of being productive in order to stay calm—they are more likely to make hasty decisions and less likely to reconsider an unwise choice. The "need for closure introduces a bias into the judgmental process," a team of researchers wrote in *Political Psychology* in 2003. A high need for closure has been shown to trigger close-mindedness, authoritarian impulses, and a preference for conflict over cooperation. Individuals with a high need for closure "may display considerable cognitive impatience or impulsivity: They may 'leap' to judgment on the basis of inconclusive evidence and exhibit rigidity of thought and reluctance to entertain views different from their own," the authors of the need for closure scale, Arie Kruglanski and Donna Webster, wrote in 1996.

Put differently, an instinct for decisiveness is great—until it's not. When people rush toward decisions simply because it makes them *feel* like they are getting something done, missteps are more likely to occur.

Researchers describe the need for closure as having multiple components. There is the need to "seize" a goal, as well as a separate urge to "freeze" on an objective once it has been selected. Decisive people have an instinct to "seize" on a choice when it meets a minimum threshold of acceptability. This is a useful impulse, because it helps us commit to projects rather than endlessly debating questions or second-guessing ourselves into a state of paralysis.

However, if our urge for closure is *too* strong, we "freeze" on our goals and yearn to grab that feeling of productivity at the expense of common sense. "Individuals with a high need for cognitive closure may deny, reinterpret or suppress information inconsistent with the preconceptions on which they are 'frozen,'" the *Political Psychology* researchers wrote. When we're overly focused on feeling productive, we become blind to details that should give us pause.

It feels good to achieve closure. Sometimes, though, we become unwilling to sacrifice that sensation even when it's clear we're making a mistake.

● ● ●

On October 1, 1973, six months after Zeira predicted that the odds of war were "very low"—and five days before Yom Kippur, the holiest day in the Jewish calendar—a young Israeli intelligence officer named Binyamin Siman-Tov sent his commanders in Tel Aviv a warning: He was receiving reports from the Sinai that large numbers of Egyptian convoys were arriving at night. Egypt's military was digging up minefields they had installed along the border, making it easier for them to move material across the canal. There were stockpiles of boats and bridge-making supplies on the Egyptian side of the border. It was the largest buildup of equipment that soldiers on the front lines had seen.

Zeira had received a number of reports like this in the preceding week, but they didn't cause him much concern. Remember the concept, he counseled his lieutenants: Egypt still didn't have enough planes or missiles to defeat Israel. And besides, Zeira had other things to focus on, most notably the cultural transformation he was pushing through the Directorate of Military Intelligence. In the midst of remaking the military's approach to threat analysis, Zeira was also ridding his agency of its propensity for endless debate. Henceforth, he had declared, intelligence officers would be evaluated on the clarity of their recommendations. Both Zeira and his chief lieutenant "lacked the patience for long and open discussions and regarded them as 'bullshit,'" the historians Uri Bar-Joseph and Abraham Rabinovich wrote. Zeira would "humiliate officers who, in his opinion, came unprepared for meetings. At least once he was heard to say that those officers who estimated in spring 1973 that a war was likely should not expect promotion." Though internal debates were tolerated to a point, "once an estimate was formulated everyone was committed to it and no one was allowed to express a different estimate outside the organization."

The Directorate had to lead by example, Zeira declared. He had been appointed to provide answers, not prolong debates. When one of Zeira's subordinates, concerned about the latest reports of Egyptian troop movements, asked to mobilize a handful of reservists to help analyze what was going on, he received a phone call. "Yoel, listen well," Zeira told the memo writer. "It is intelligence's job to safeguard the nation's nerves, not to drive the public crazy." The request was denied.

On October 2 and 3, 1973, sightings of Egyptian troops increased. Then came word of activity on the border with Syria. Alarmed, the prime minister called another meeting. Zeira's division, once again, counseled that there was no reason to be concerned: Egypt and Syria had weak air forces; they had no missiles capable of hitting Tel Aviv. This time, the military experts who had disagreed with Zeira six months earlier followed his lead. "I don't see a concrete danger in the near future," one general told the prime minister. Meir was troubled before the meeting, she later recounted in her memoirs, but the intelligence estimate eased her mind. She had chosen the right officials to bring the nation much-needed relief.

Seventy-two hours after Binyamin Siman-Tov submitted his report, Israel's intelligence analysts learned that the Soviet Union had started an emergency airlift of Soviet advisers and their families out of Syria and Egypt. Intercepted telephone calls among Russian families revealed they had been ordered to hurry to the airport. Aerial photographs showed more tanks, artillery, and air-defense guns massing along the Suez Canal and in the Syrian-controlled portions of the Golan Heights.

On the morning of Friday, October 5, four days after Siman-Tov's report, a group of Israel's top military commanders, including Zeira, gathered in the office of Defense Minister Moshe Dayan. The hero of the Six-Day War was upset. The Egyptians had positioned 1,100 pieces of artillery along the Suez, and air reconnaissance showed massive numbers of troops. "You people don't take the Arabs seriously enough," Dayan said. The chief of staff of the Israeli Defense Forces agreed. Earlier that morning he had ordered the army to its highest alert since 1967.

But Zeira had another explanation for the troop movements: The Egyptians were preparing their defenses in case Israel launched an invasion of its own. There were no new fighter jets in Egypt, he said. No Scud missiles. Arab leaders knew striking at Israel would be suicide. "I don't see either the Egyptians or the Syrians attacking," Zeira said.

Afterward, the meeting moved into the prime minister's office. She asked for an update. The military's chief of staff, aware that mobilizing Israel's reservists on the holiest Jewish holiday would draw fierce criticism, said, "I still think that they're not going to attack, but we have no hard information."

Then Zeira spoke. Concerns about the Egyptians and Syrians attacking, he said, were "absolutely unreasonable." He even had a logical reason for the evacuation of Soviet advisers. "Maybe the Russians think the Arabs are going to attack because they don't understand them well," he said, but the Israelis knew their neighbors better than that. Later that day, when Israeli generals briefed the prime minister's cabinet, Zeira reiterated that he believed there was a "low probability" of war. What they were seeing were defensive preparations or a military exercise, Zeira argued. Arab leaders weren't irrational.

Having seized on an answer—that Egypt and Syria knew they couldn't win, and therefore wouldn't attack—Zeira was frozen, unwilling to reconsider the question. His goal of disciplined decision making had been satisfied.

The next morning was the first full day of Yom Kippur.

Before daybreak, the head of the Mossad telephoned his colleagues to say that a well-connected source had told him Egypt would invade by nightfall. The message was delivered to the prime minister as well as Dayan and the military's chief of staff. They all rushed to their offices as the sun rose. The odds of war, they believed, had just shifted.

As Yom Kippur prayers began, Israel's streets were quiet. Families were gathered in homes and synagogues. Shortly after ten o'clock, a full six days after enemy forces had started massing along Israeli borders, the military finally issued a partial call-up of reserves. Inside houses of worship, rabbis read hastily delivered lists with the names of people who needed to report for duty. By then, Egypt and Syria had been moving tanks and artillery into offensive range for weeks, but this was the first public hint that trouble might be near. At that moment, there were more than 150,000 enemy soldiers along Israel's borders, ready to attack from two directions, and another half million soldiers waiting to follow the initial waves. Egypt and Syria had been coordinating their invasion plans for months.

When confidential documents from that period were released decades later, they revealed that Egypt's president had assumed Israel knew what he was doing. How else could they interpret all the men and matériel being moved to the border?

Meir called an emergency meeting of her cabinet for noon. "She was pale and her eyes were downcast," *The Times of Israel* wrote in a reconstruction of that day. "Her hair, normally neatly combed and pulled back, was disheveled and she looked like she had not shut her eyes all night. . . . She began with a detailed report of events over the past few days—the Arab deployment on the borders that had suddenly taken on ominous color, the hasty evacuation of the families of Soviet advisers from Egypt and Syria, the air photos, the insistence by military intelligence that there would be no war despite mounting evidence to the contrary." Meir announced her conclusion: An invasion of Israel was likely to occur, maybe as soon as within the next six hours.

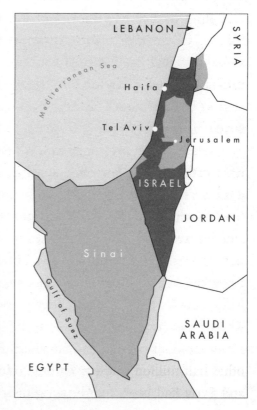

"The ministers were stunned," *The Times of Israel* reported. "They had not been made privy to the Arab buildup. Furthermore, they had been told for years that even in a worst-case situation, military intelligence would provide at least a 48-hour warning to call up the reserves before war broke out." Now they were being informed that a two-front war would occur in less than six hours. The reserves were only partially mobilized—and because of the holiday, it was unclear how quickly troops would be able to get to the front.

The attack came even sooner than Meir expected. Two hours after the cabinet meeting started, the first of ten thousand Egyptian shells began falling on the Sinai; at four P.M., twenty-three thousand Egyptian soldiers crossed the Suez in the first wave of attack. By the end of the day, enemy forces were two miles into Israeli territory. They had killed five hundred Israeli soldiers and were rapidly advancing toward the Israeli towns of Yamit and Avshalom, as well as an Israeli air force base. Meanwhile, on the other side of the country, Syria struck simultaneously, attacking the Golan Heights with soldiers, planes, and tanks.

Over the next twenty-four hours, Egypt and Syria pushed deeper into the Sinai and the Golan as Israel scrambled to respond. Over a hundred thousand enemy troops were inside Israel's territory. It took three days to halt the Egyptian advance, and two days to organize a counterstrike against Syria. Eventually, Israel's superior firepower asserted itself. Israeli soldiers drove the Syrian army back toward the border, forcing the retreating army to leave behind 1,000 of its 1,500 tanks. A few days later, Israeli Defense Forces began shelling the outskirts of Damascus.

Then Egyptian president Anwar Sadat, hoping to take more territory in the Sinai, launched a risky offensive to capture two strategic passes deep inside the peninsula. The gamble failed. Israeli forces pushed the Egyptians backward. On October 15, nine days after Egypt invaded, Israel crossed the Suez Canal and began taking Egyptian land. Within the week, Egypt's Third Army, located along the banks of the Suez, was encircled, cut off by the Israelis from supplies and reinforcements. The Second Army, in the north, was almost completely surrounded as well. In the face of defeat, President Sadat demanded a cease-fire, and American and Soviet leaders pressured Israel to agree. Fighting stopped in late October, and the war formally ended on January 18, 1974. Israel had repelled the invasion, but at a huge cost. More than ten thousand Israelis were killed or wounded. As many as thirty thousand Egyptians and Syrians are estimated to have died.

"Something of ours was destroyed on Yom Kippur last year," an Israeli newspaper wrote on the first anniversary of the war. "The state was saved, true, but our faith was fractured, our trust damaged, our hearts deeply gouged, and an entire generation was nearly lost."

"Even a quarter century later the Yom Kippur war remains the most traumatic phase in Israel's history," the historian P. R. Kumaraswamy wrote. Today, the psychological scars of the invasion are still profound.

Zeira had set out with a goal of alleviating public anxiety, and the government had followed his lead. But in their eagerness to provide confident answers, to make decisive judgments and avoid ambiguity, those leaders had almost cost Israel its life.

III.

Fifteen years later and half a world away, General Electric, one of the largest companies on earth, was thinking about very different kinds of goals when executives contacted an organizational psychologist from the University of Southern California and asked him for help figuring out why some factories had gone awry.

It was the late 1980s and GE was the second most valuable company in America, just behind Exxon. GE manufactured everything from lightbulbs to jet engines, refrigerators to railway cars, and through its ownership of NBC, was in millions of homes with iconic shows such as *Cheers, The Cosby Show,* and *L.A. Law.* The company employed over 220,000 people, more than many U.S. cities had residents. One of the reasons GE was so successful, its executives boasted, was that it was so good at choosing goals.

In the 1940s, GE had formalized a corporate goal-setting system that would eventually become a model around the world. By the

SMART GOALS

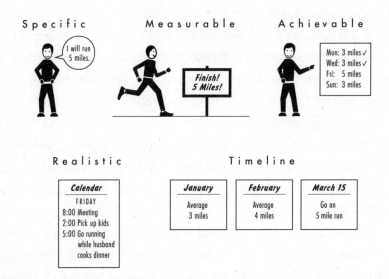

1960s, every GE employee was required to write out their objectives for the year in a letter to their manager. "Simply put," historians at Harvard Business School wrote in 2011, "the manager's letter required a job holder to write a letter to his or her superior indicating what the goals for the next time frame were, how the goals would be met, and what standards were to be expected. When the superior accepted this letter—usually after editing and discussion—it became the work 'contract.'"

By the 1980s, this system had evolved into a system of so-called SMART goals that every division and manager were expected to describe each quarter. These objectives had to be specific, measurable, achievable, realistic, and based on a timeline. In other words, they had to be provably within reach and described in a way that suggested a concrete plan.

If a goal didn't meet the SMART criteria, a manager had to resubmit a memo detailing their aims, again and again, until it was approved by upper management. "It was about getting concrete," said William Conaty, who retired as GE's head of human resources in 2007. "Your manager was always saying, what's the specifics? What's the timeline? Prove to me this is realistic. The system worked because by the time we were done, you knew pretty clearly how things were going to unfold."

The SMART mindset spread throughout GE's culture. There were SMART charts to help midlevel managers describe monthly goals and SMART worksheets to turn personal objectives into action plans. And the company's belief that SMART goals would work was rooted in good science.

In the 1970s, a pair of university psychologists named Edwin Locke and Gary Latham had helped develop the SMART criteria through experiments scrutinizing the best way to set goals. In one experiment Latham conducted in 1975, researchers approached forty-five of the most experienced and productive typists at a large corporation and measured how fast they produced text. The typists knew they were among the best in the company, but they had never measured how quickly they typed. The researchers found that, on average, each worker produced ninety-five lines of typewritten output per hour.

Then the researchers gave each typist a specific goal based on their previous performance—such as ninety-eight lines per hour— and showed the typists a system for easily measuring their hourly output. The researchers also had a conversation with each typist to make sure the goal was realistic—and to adjust it if necessary— and they discussed what changes were required to make the objective achievable. They came up with a timeline for each person. The conversation didn't take long—say, fifteen minutes per person—but afterward each typist knew exactly what to do and how to measure success. Each of them, put differently, had a SMART goal.

Some of the researchers' colleagues said they didn't believe this would have an impact on the typists' performance. All the typists were professionals with years of experience. A fifteen-minute conversation should not make much of a difference to someone who has been typing eight hours a day for two decades.

But one week later, when the researchers measured typing speeds again, they found that the workers, on average, were completing 103 lines per hour. Another week later: 112 lines. Most of the typists had blown past the goals they had set. The researchers worried the workers were just trying to impress them, so they came back again, three months later, and quietly measured everyone's performance once more. They were typing just as fast, and some had gotten even faster.

"Some 400 laboratory and field studies [show] that specific, high goals lead to a higher level of task performance than do easy goals or vague, abstract goals such as the exhortation to 'do one's best,'" Locke and Latham wrote in 2006 in a review of goal-setting studies. In particular, objectives like SMART goals often unlock a potential that people don't even realize they possess. The reason, in part, is because goal-setting processes like the SMART system force people to translate vague aspirations into concrete plans. The process of making a goal specific and proving it is achievable involves figuring out the steps it requires—or shifting that goal slightly, if your initial aims turn out to be unrealistic. Coming up with a timeline and a way to measure success forces a discipline onto the process that good intentions can't match.

"Making yourself break a goal into its SMART components is the difference between hoping something comes true and figuring out how to do it," Latham told me.

GE's chief executive Jack Welch had long claimed that his insistence on SMART goals was one of the reasons the company's stock had more than tripled in eight years. But forcing people to detail their goals with such specificity didn't mean every part of the company ran smoothly. Some divisions, despite setting SMART goals, never seemed to excel—or they would flip-flop from profits to losses, or seem to be growing and then suddenly fall apart. In the late 1980s, executives became particularly concerned about two divisions—a nuclear equipment manufacturer in North Carolina and a jet engine plant in Massachusetts—that had once been among the company's top performers but were now limping along.

Executives initially suspected those divisions simply needed better-defined objectives, so factory managers were asked to prepare memo after memo describing increasingly specific goals. Their responses were detailed, precise, and realistic. They met every SMART criterion.

And yet, profits still fell.

So a group of GE's internal consultants visited the nuclear factory in Wilmington, North Carolina. They asked employees to walk them through their weekly, monthly, and quarterly goals. One plant executive explained that his SMART objective was to prevent anti-nuclear protesters from harassing workers as they entered the plant, because he felt it eroded morale. He had come up with a SMART plan to build a fence. The goal was specific and reasonable (the fence would be fifty feet long and nine feet high), it had a timeline (it would be done by February), and it was achievable (they had a contractor ready to go).

Next, the consultants went to the jet engine factory in Lynn, Massachusetts, and interviewed, among others, an administrative assistant who told them her SMART goal was ordering the factory's office supplies. She showed them a SMART chart with specific aims ("order staplers, pens and desk calendars") that were measurable ("by June"), as well as achievable, realistic, and had a timeline ("Place order on February 1. Request update on March 15.").

Many of the SMART goals the consultants found inside the factories were just as detailed—and just as trivial. Workers spent hours making sure their objectives satisfied every SMART criterion, but spent much less time making sure the goals were worth pursuing in the first place. The nuclear factory's security guards had written extensive memos on the goal of theft prevention and had come up with a plan that "basically consisted of searching everyone's bags every time they entered or exited the plant, which caused huge delays," said Brian Butler, one of the consultants. "It might have stopped thefts, but it also destroyed the factory's productivity because everyone started leaving earlier each day so they could get home at a decent hour." Even the plants' senior executives, the consultants found, had fallen prey to an obsession with achievable but inconsequential goals, and were focused on unimportant short-term objectives rather than more ambitious plans.

When the consultants asked employees how they felt about GE's emphasis on SMART goals, they expected to hear complaints about the onerous bureaucracy. They anticipated people would say they *wanted* to think bigger, but were hamstrung by the incessant SMART demands. Instead, employees said they loved the SMART system. The administrative assistant who ordered office supplies said fulfilling those goals gave her a real sense of accomplishment. Sometimes, she said, she would write a SMART memo for a task she had already completed and then put it into her "Done" folder. It made her feel so good.

Researchers who have studied SMART goals and other structured methods of choosing objectives say this isn't unusual. Such systems, though useful, can sometimes trigger our need for closure in counterproductive ways. Aims such as SMART goals "can cause [a] person to have tunnel vision, to focus more on expanding effort to get immediate results," Locke and Latham wrote in 1990. Experiments have shown that people with SMART goals are more likely to seize on the easiest tasks, to become obsessed with finishing projects, and to freeze on priorities once a goal has been set. "You get into this mindset where crossing things off your to-do list becomes more important than asking yourself if you're doing the right things," said Latham.

GE's executives weren't sure how to help the nuclear and jet engine factories. So in 1989, they asked a professor named Steve Kerr, the dean of faculty at the University of Southern California business school, for help. Kerr was an expert in the psychology of goal setting, and he began by interviewing employees inside the nuclear factory. "A lot of these people were really demoralized," he said. "They had gone into nuclear energy because they wanted to change the world. Then Three Mile Island and Chernobyl happened, and the industry was getting protested every day and completely brutalized in the press." Setting short-term goals and achieving them, the plant's workers and executives told Kerr, was one of the few things they could feel good about at work.

The only way to improve performance at the nuclear factory, Kerr thought, was to find a way to shake people out of their focus on short-term objectives. GE had recently started a series of meetings among top executives called "Work-Outs" that were designed to encourage people to think about bigger ambitions and more long-term plans. Kerr helped expand those meetings to factories' rank-and-file.

The rules at Work-Outs were simple: Employees could suggest any goal they thought GE *ought* to be pursuing. There were no SMART charts or memos. "The concept was that nothing was off-limits," Kerr told me. Managers had to approve or deny each suggestion quickly, often right away, and "we wanted to make it easy to say yes," said Kerr. "We thought if we could get people to identify the ambition first, and then figure out the plan afterward, it would encourage bigger thinking." If an idea seemed half-baked, Kerr said, a manager should "say yes, because even if the proposal is no better than what you're doing now, with the group's energy behind it, the plan will turn out great." Only after a goal was approved would everyone begin the formal process of determining how to make it realistic and achievable and all the other SMART criteria.

At a Work-Out inside the engine factory in Massachusetts, one worker told his bosses they were making a mistake by outsourcing construction of protective shields for their grinding machines. The factory could make them in-house for half the cost, he said. Then he unfurled a piece of butcher paper covered with scribbled blueprints.

There was nothing SMART about the man's proposal. It was unclear if it was realistic or achievable, or what measurements to apply. But when the factory's top manager looked at the butcher paper, he said, "I guess we'll try it out."

Four months later, after the blueprints had been professionally redrawn and the plan transformed into a series of SMART goals, the first prototype was installed. It cost $16,000—more than 80 percent less than the outsourced bid. The factory saved $200,000 that year on ideas proposed at the Work-Out. "Everybody gets caught up in this tremendous rush of adrenaline," a team leader at the plant, Bill DiMaio, said. "The ideas that people come up with are so encouraging, it's unbelievable. These people get psyched. All their ideas are fair game."

Then Kerr helped take the Work-Out program company-wide. By 1994, every GE employee within GE had participated in at least one Work-Out. As profits and productivity rose, executives at other companies began imitating the Work-Out system inside their own firms. By 1995, there were hundreds of companies conducting Work-Outs. Kerr joined GE full-time in 1994 and eventually became the company's "chief learning officer."

"The Work-Outs were successful because they balanced the psychological influence of immediate goals with the freedom to think about bigger things," said Kerr. "That's critical. People respond to the conditions around them. If you're being constantly told to focus on achievable results, you're only going to think of achievable goals. You're not going to dream big."

Work-Outs, however, weren't perfect. They took an entire day of everyone's time and usually meant the plant had to slow down production so that workers could all attend the meetings. It was something a division or plant could do once or twice a year, at most. And though the Work-Outs left everyone feeling excited and hungry for change, the effects were frequently short-lived. A week later, everyone was back at their old jobs and, often, their old ways of thinking.

Kerr and his colleagues wanted to foster perpetual ambitions. How, they wondered, do you get people to think expansively all the time?

IV.

In 1993, twelve years after becoming chief executive of General Electric, Jack Welch traveled to Tokyo and, while touring a factory that made medical testing equipment, heard a story about Japan's railway system.

In the 1950s, during the lingering wake of the devastation of the Second World War, Japan was intensely focused on growing the nation's economy. A large portion of the country's population lived in or between the cities of Tokyo and Osaka, which were separated by just 320 miles of train track. Every day, tens of thousands of people traveled between the cities. Vast amounts of raw industrial materials were transported on those rail lines. But the Japanese topography was so mountainous and the railway system so outdated that the trip could take as long as twenty hours. So, in 1955, the head of the Japanese railway system issued a challenge to the nation's finest engineers: invent a faster train.

Six months later, a team unveiled a prototype locomotive capable of going 65 miles per hour—a speed that, at the time, made it among the fastest passenger trains in the world. Not good enough, the head of the railway system said. He wanted 120 miles per hour.

The engineers explained that was not realistic. At those speeds, if a train turned too sharply, the centrifugal force would derail the cars. Seventy miles an hour was more realistic—perhaps 75. Any faster and the trains would crash.

Why do the trains need to turn? the railway head asked.

There were numerous mountains between the cities, the engineers replied.

Why not make tunnels, then?

The labor required to tunnel through that much territory could equal the cost of rebuilding Tokyo after World War II.

Three months later, the engineers unveiled an engine capable of going 75 miles per hour. The railway chief lambasted the designs. Seventy-five miles per hour, he said, had no chance of transforming the nation. Incremental improvements would only yield incremental economic growth. The only way to overhaul the nation's transportation system was to rebuild every aspect of how trains functioned.

Over the next two years, the engineers experimented: They designed train cars that each had their own motors. They rebuilt gears so they meshed with less friction. They discovered that their new cars were too heavy for Japan's existing tracks, and so they reinforced the rails, which had the added bonus of increasing stability, which added another half mile per hour to cars' speed. There were hundreds of innovations, large and small, that each made the trains a little bit faster than before.

In 1964, the Tōkaidō Shinkansen, the world's first bullet train, left Tokyo along continuously welded rails that passed through tunnels cut into Japan's mountains. It completed its inaugural trip in three hours and fifty-eight minutes, at an average speed of 120 miles per hour. Hundreds of spectators had waited overnight to see the train arrive in Osaka. Soon other bullet trains were running to other Japanese cities, helping fuel a dizzying economic expansion. The development of the bullet train, according to a 2014 study, was critical in spurring Japan's growth well into the 1980s. And within a decade of that innovation, the technologies developed in Japan had given birth to high-speed rail projects in France, Germany, and Australia, and had revolutionized industrial design around the world.

For Jack Welch, this story was a revelation. What GE needed, he told Kerr when he got home from Japan, was a similar outlook, an institutional commitment to audacious goals. Going forward, every executive and department, in addition to delivering specific and achievable and timely objectives, would *also* have to identify a stretch goal—an aim so ambitious that managers couldn't describe, at least initially, how they would achieve it. Everyone, Welch said, had to partake in "bullet train thinking."

In a 1993 letter to shareholders, the chief executive explained that "stretch is a concept that would have produced smirks, if not laughter, in the GE of three or four years ago, because it essentially means using dreams to set business targets—with no real idea of how to get there. If you do know how to get there—it's not a stretch target."

Six months after Welch's trip to Japan, every division at GE had a stretch goal. The division manufacturing airplane engines, for instance, announced they would reduce the number of defects in finished engines by 25 percent. To be honest, the division's managers figured they could hit that target pretty easily. Almost all the defects they found on engines were small, cosmetic issues, such as a slightly misaligned cable or unimportant scratches. Anything more serious was corrected before the engine was shipped. If they hired more quality assurance employees, managers figured, they could reduce cosmetic defects with little effort.

Welch agreed that reducing defects was a wise goal.

Then he told them to cut errors by 70 percent.

That's ridiculous, managers said. Manufacturing engines was such a complicated affair—each one weighed five tons and had more than ten thousand parts—that there was no way they could achieve a 70 percent reduction.

They had three years, Welch said.

The division's managers started panicking—and then began analyzing every error that had been recorded in the previous twelve months. Simply hiring more quality assurance workers, they quickly realized, wouldn't do the trick. The only way to reduce errors by 70 percent was to make every single employee, in effect, a quality assurance auditor. Everyone had to take responsibility for catching mistakes. But most factory workers didn't know enough about the engines to identify every small defect as it occurred. The only solution, managers decided, was a massive retraining effort.

Except that didn't really work, either. Even after nine months of retraining, the error rate had fallen by only 50 percent. So managers started hiring workers with more technical backgrounds, the kind of people who knew what an engine ought to look like and, therefore, could more easily spot what was amiss. The GE factory manufacturing CF6 engines in Durham, North Carolina, determined that the best way to find the right employees was to hire only candidates with FAA certification in engine manufacturing. Such workers, however, were already in high demand at other plants. So to attract them, managers said employees could have more autonomy. They could schedule their own shifts and organize teams however they wanted. That required the plant to do away with centralized scheduling. Teams had to self-organize and figure out their own workflow.

Welch had given his aircraft manufacturing division a stretch goal of reducing errors by 70 percent, an objective so audacious the only way to go about it was to change nearly everything about (a) how workers were trained, (b) which workers were hired, and (c) how the factory ran. By the time they were done, the Durham plant's managers had collapsed organizational charts, remade job duties, and overhauled how they interviewed candidates, because they needed people with better team skills and more flexible mindsets. In other words, Welch's stretch goal set off a chain reaction that remade how engines were manufactured in ways no one had imagined. By 1999, the number of defects per engine had fallen by 75 percent and the company had gone thirty-eight months without missing a single delivery, a record. The cost of manufacturing had dropped by 10 percent every year. No SMART goal would have done that.

Numerous academic studies have examined the impact of stretch goals, and have consistently found that forcing people to commit to ambitious, seemingly out-of-reach objectives can spark outsized jumps in innovation and productivity. A 1997 study of Motorola, for instance, found that the time it took engineers to develop new products fell tenfold after the company mandated stretch goals throughout the firm. A study of 3M said stretch goals helped spur such inventions as Scotch tape and Thinsulate. Stretch goals transformed Union Pacific, Texas Instruments, and public schools in Washington, D.C., and Los Angeles. Surveys of people who have lost large amounts of weight or have become marathon runners later in life have found that stretch goals are often integral to their success.

Stretch goals "serve as jolting events that disrupt complacency and promote new ways of thinking," a group of researchers wrote in *Academy of Management Review* business journal in 2011. "By forcing a substantial elevation in collective aspirations, stretch goals can shift attention to possible new futures and perhaps spark increased energy in the organization. They thus can prompt exploratory learning through experimentation, innovation, broad search, or playfulness."

There is an important caveat to the power of stretch goals, however. Studies show that if a stretch goal is audacious, it can spark innovation. It can also cause panic and convince people that success is impossible because the goal is *too* big. There is a fine line between an ambition that helps people achieve something amazing and one that crushes morale. For a stretch goal to inspire, it often needs to be paired with something like the SMART system.

The reason why we need both stretch goals and SMART goals is that audaciousness, on its own, can be terrifying. It's often not clear how to start on a stretch goal. And so, for a stretch goal to become more than just an aspiration, we need a disciplined mindset to show us how to turn a far-off objective into a series of realistic short-term aims. People who know how to build SMART goals have often been habituated into cultures where big objectives can be broken into manageable parts, and so when they encounter seemingly outsized ambitions, they know what to do. Stretch goals, paired with SMART thinking, can help put the impossible within reach.

In one experiment conducted at Duke University, for instance, varsity athletes were asked to run around a track and, when signaled, get as close as possible to a finish line 200 meters away within ten seconds. The runners in the study all knew, simply by looking at the distance they were being asked to cover, that the goal was absurd. No person has ever run anything close to 200 meters in ten seconds. The athletes made it 59.6 meters, on average, during their sprint.

A few days later, those same participants were presented with the same task, but this time the finish line was only 100 meters away. The goal was still audacious—but it was within the realm of possibility. (Usain Bolt ran 100 meters in 9.58 seconds in 2009.) During this trial, the runners made it, on average, 63.1 meters in ten seconds—"a large difference by track and field standards," the researchers noted.

This difference in performance was explained by the fact that the shorter distance, while still challenging, lent itself to the kind of methodical planning and mental models that experienced runners are accustomed to using. The shorter distance, in other words, allowed the runners to participate in the athletic equivalent of breaking a stretch goal into SMART components. "All runners in our sample engaged in regular workouts," the researchers wrote, and so when confronted with running 100 meters in ten seconds, they knew how to wrestle with the task. They broke it into pieces and treated it like they would other sprints. They started strong, and paced off other runners, and then pushed themselves as hard as possible in the final seconds. But when they were confronted with running 200 meters in ten seconds, there was no practical approach. There was no way to break the problem into manageable parts. There were no SMART criteria they could apply. It was simply impossible.

Experiments at the University of Waterloo, the University of Melbourne, and elsewhere show similar results: Stretch goals can spark remarkable innovations, but only when people have a system for breaking them into concrete plans.

This lesson can extend to even the most mundane aspects of life. Take, for instance, to-do lists. "To-do lists are great if you use them correctly," Timothy Pychyl, a psychologist at Carleton University, told me. "But when people say things like 'I sometimes write down easy items I can cross off right away, because it makes me feel good,' that's exactly the *wrong* way to create a to-do list. That signals you're using it for mood repair, rather than to become productive."

The problem with many to-do lists is that when we write down a series of short-term objectives, we are, in effect, allowing our brains to seize on the sense of satisfaction that each task will deliver. We are encouraging our need for closure and our tendency to freeze on a goal without asking if it's the right aim. The result is that we spend hours answering unimportant emails instead of writing a big, thoughtful memo—because it feels so satisfying to clean out our in-box.

At first glance, it might seem like the solution is creating to-do lists filled solely with stretch goals. But we all know that merely writing down grand aspirations doesn't guarantee we will achieve them. In fact, studies show that if you're confronted with a list of only far-reaching objectives, you're more likely to get discouraged and turn away.

So one solution is writing to-do lists that pair stretch goals and SMART goals. Come up with a menu of your biggest ambitions. Dream big and stretch. Describe the goals that, at first glance, seem impossible, such as starting a company or running a marathon.

Then choose one aim and start breaking it into short-term, concrete steps. Ask yourself: What realistic progress can you make in the next day, week, month? How many miles can you realistically run tomorrow and over the next three weeks? What are the specific, short-term steps along the path to bigger success? What timeline

THE GOAL-SETTING FLOWCHART

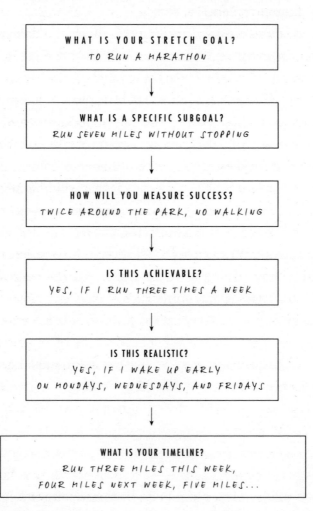

WHAT IS YOUR STRETCH GOAL?
TO RUN A MARATHON

WHAT IS A SPECIFIC SUBGOAL?
RUN SEVEN MILES WITHOUT STOPPING

HOW WILL YOU MEASURE SUCCESS?
TWICE AROUND THE PARK, NO WALKING

IS THIS ACHIEVABLE?
YES, IF I RUN THREE TIMES A WEEK

IS THIS REALISTIC?
YES, IF I WAKE UP EARLY
ON MONDAYS, WEDNESDAYS, AND FRIDAYS

WHAT IS YOUR TIMELINE?
RUN THREE MILES THIS WEEK,
FOUR MILES NEXT WEEK, FIVE MILES...

makes sense? Will you open your store in six months or a year? How will you measure your progress? Within psychology, these smaller ambitions are known as "proximal goals," and repeated studies have shown that breaking a big ambition into proximal goals makes the large objective more likely to occur.

When Pychyl writes a to-do list, for instance, he starts by putting a stretch goal—such as "conduct research that explains goal/neurology interface"—at the top of the page. Underneath comes the nitty-gritty: the small tasks that tell him precisely what to do. "Specific: Download grant application. Timeline: By tomorrow."

"That way, I'm constantly telling myself what to do next, but I'm also reminded of my larger ambition so I don't get stuck in the weeds of doing things simply to make myself feel good," Pychyl said.

In short, we need stretch *and* SMART goals. It doesn't matter if you call them by those names. It's not important if your proximal goals fulfill every SMART criterion. What matters is having a large ambition and a system for figuring out how to make it into a concrete and realistic plan. Then, as you check the little things off your to-do list, you'll move ever closer to what really matters. You'll keep your eyes on what's both wise and SMART.

"I had no idea how what we were doing would affect the rest of the world," Kerr told me. GE's embrace of SMART and stretch goals has been analyzed in academic studies and psychology textbooks; the firm's system has been imitated throughout corporate America. "We proved you can change how people act by asking them to think about goals differently," said Kerr. "Once you know how to do that, you can get pretty much anything done."

V.

Twenty-seven days after fighting concluded in the Yom Kippur War, the Israeli parliament established a national committee of inquiry to examine why the nation had been so dangerously unprepared. Officials met for 140 sessions and heard testimony from fifty-eight witnesses, including Prime Minister Golda Meir, Defense Minister Moshe Dayan, and the head of the Directorate of Military Intelligence, Eli Zeira.

"In the days that preceded the Yom Kippur war, the Research Division of Military Intelligence had plenty of warning indicators," investigators concluded. There was no justification for Israel to have been caught off guard. Zeira and his colleagues had ignored obvious signs of danger. They had dissuaded other leaders from following their instincts. These mistakes were not made out of malice, investigators said, but because Zeira and his staff had become so obsessed with avoiding unnecessary panic and making firm decisions that they lost sight of their most important objective: keeping Israelis safe.

Prime Minister Meir resigned one week after the government's report was released. Moshe Dayan, the onetime hero, was hounded by critics until his death six years later. And Zeira was relieved of his position and forced to resign from government service.

Zeira's failings in the run-up to the Yom Kippur War illustrate one final lesson regarding how goals function and influence our psychology. He, in fact, was using both stretch and SMART goals when he convinced the nation's leaders to ignore obvious signs of war. He had clear and grand ambitions to end the cycle of anxiety plaguing Israelis; he knew that his big aim was to stop the endless debates and second-guessing. And his methods for breaking those larger goals into smaller pieces involved finding proximal goals that were specific, measurable, achievable, and realistic, and that occurred according to a timeline. He remade his agency in a deliberate, step-by-step manner. He did everything that psychologists like Latham and Locke have said we ought to do in order to achieve both big and small goals.

Yet Zeira's craving for closure and his intolerance for revisiting questions once they were answered are among the biggest reasons why Israel failed to anticipate the attacks. Zeira is an example of how stretch and SMART goals, on their own, sometimes aren't enough. In addition to having audacious ambitions and plans that are thorough, we still need, occasionally, to step outside the day-to-day and consider if we're moving toward goals that make sense. We still need to *think*.

On October 6, 2013, the fortieth anniversary of the Yom Kippur War, Eli Zeira addressed an audience of national security scholars in

Tel Aviv. He was eighty-five years old and his gait was a bit unsteady as he walked onto the stage. He spoke haltingly from handwritten notes. He had come to defend himself, he said. Mistakes had been made, but not just by him. Everyone had learned they needed to be more careful and less certain. They were all to blame.

A former colleague in the audience began heckling him.

"You are telling us fairy tales!" the man shouted. "You are lying!"

"This is not a field court marshal," Zeira replied. The war was not his fault alone, he said. No one had been willing to stare the most terrifying possibility—a full-scale invasion—in the face.

But then, in a moment of reflection, Zeira conceded that he had made an error. He had ignored the seemingly impossible. He hadn't thought through all the alternatives as deeply as he should.

"I usually had a note in my pocket," he told the audience, "and on that little note, it said, *and if not?*'" The note was a talisman, a reminder that the desire to get something done, to be decisive, can also be a weakness. The note was supposed to prompt him to ask bigger questions.

But in the days before the Yom Kippur War, "I didn't read that little note," Zeira said. "That was my mistake."

5

MANAGING OTHERS

Solving a Kidnapping with Lean and Agile Thinking and a Culture of Trust

Frank Janssen had just returned home from a bike ride when he heard a knock at the front door. It was a sunny Saturday morning; there were kids playing soccer a few blocks away. When Janssen looked out the window, he saw a woman holding a clipboard and two men dressed in khakis and button-down shirts. Perhaps they were conducting a survey? Or they were religious missionaries? Janssen didn't know why they were on his doorstep, but he hoped it wouldn't take more than a moment to shoo them away.

When he opened the door, however, the men pushed their way inside. One of them grabbed Janssen, shoved him against the wall, and then threw him to the floor. He pulled a gun from his waistband and slammed the barrel across Janssen's face. The other man pressed a stun gun against Janssen's torso and pulled the trigger, momentarily paralyzing the sixty-three-year-old. Then they bound his hands with a plastic zip tie and carried him outside, into the backseat of a silver Nissan waiting in the driveway. The two men sat on either side of Janssen while the woman sat up front, next to the driver. As Janssen slowly regained control of his body, he began shoving his attackers. They pushed him to the floor and applied the stun gun again. The car backed onto the street and headed west, past the field where kids were playing soccer. One of the assailants draped a blanket over Janssen's body. The vehicle turned onto a freeway and slipped into southbound traffic.*

Janssen's wife came home about an hour later and found the house empty and the front door ajar. Frank's bike was propped against the garage. Maybe he had gone for a walk? An hour later, with no sign of him, his wife grew concerned. She searched the entryway, thinking he might have left a note. On the doorstep, she saw a few drops of blood. Panicked, she walked toward the driveway and found more blood outside. She phoned her daughter, who told her to call the police.

Her husband, she explained to officers, was a consultant at a firm specializing in national security. Soon her home was surrounded by police cruisers and yellow crime-scene tape. Black SUVs pulled up, delivering a team of FBI agents who dusted for prints and photographed indentations in the grass. For the next two days, agents pored over Janssen's cellphone records and interviewed neighbors and coworkers, but found nothing to indicate what was going on.

Then, three days after the abduction, in the middle of the night of April 7, 2014, his wife's phone buzzed. It was a series of text messages from an unfamiliar number with a New York City area code.

We have your husband, the texts read, and he is in the trunk of a car going to California. If she contacted the police, *we will send him*

---

* The Federal Bureau of Investigation was provided with summaries of this chapter. Please see the chapter's endnotes for the bureau's responses. The Janssen family did not reply to repeated attempts to seek their comments by telephone and certified mail. Details regarding this case come from court documents, interviews, and other materials specified in the endnotes. At the time of writing, the allegations of criminal activity contained in this chapter had not been adjudicated by a court of law, and thus remain allegations rather than proven facts. Please see the chapter's endnotes for further details and the responses provided by the attorneys of those implicated in this alleged crime.

back to you in 6 boxes and every chance we get we will take someone in you family to italy and torture them and kill them, we will do a drive by and gun down anybody in you family and we will throw grenades in you window.

The texts also referenced Janssen's daughter and a man named Kelvin Melton. Suddenly, things started making a bit more sense. Janssen's daughter, Colleen, was an assistant district attorney in nearby Wake Forest, and she had prosecuted Melton, a high-ranking gang member in the Bloods, a few years earlier. Colleen had successfully sent Melton to jail for the rest of his life on a charge of assault with a deadly weapon. A theory began to emerge: Government investigators suspected that the Bloods had kidnapped Frank Janssen to punish his daughter. This was revenge for putting one of their leaders behind bars.

Within hours, the police had subpoenaed the records of the phone sending the texts looking for a link with known gang members. They could tell the messages had been transmitted from Georgia, but the device was a burner, an unregistered phone bought with cash at a Walmart. There was nothing in the cellular records or purchase receipts that told investigators who owned the phone or where it was currently located.

Two days later, another text arrived from a different number, this time with an Atlanta area code. *Here is 2 picture of you husband,* it read and included photos of Janssen tied to a chair. *If you can not tell me where my things are at tomorrow i will start torchering colleen father.* None of the investigators had any idea what "things" the kidnappers were referring to. The texts also demanded that someone bring Melton, the incarcerated gang leader, a pack of cigarettes, as well as other commands. *Jefe wants his things and he needs to get another phone fast so we can finish our business and if I don't get word from him very very fast then we have problems with his people.* The police didn't know if "Jefe" referred to Melton or someone else, or why Melton would want cigarettes delivered to him since he could buy them inside the Polk Correctional Institution. More texts arrived with references to unknown people. *Now he know ah playin games,* one said. *Tell him we got franno, tell him he better find a way to tell me where my things at an get my money or we kill these people in 2 days.* Investigators were confused by these mentions of "Jefe" and "Franno," and the threats to kill multiple people even though authorities were aware of only one kidnapping victim. If this was a revenge plot, why were the kidnappers sending so many ambiguous messages? Why hadn't they made any ransom demands? One federal agent thought the kidnappers were acting as if they weren't sure what was going on themselves, as if they didn't have a plan.

The FBI asked Google to look for searches around the time of the abduction that had included Janssen's address. The computer giant reported that someone using a disposable T-Mobile phone had Googled "Colleen Janssen address," but what had come up was her parents' home, where she had once lived. A new theory emerged: The kidnappers had *intended* to kidnap Colleen as revenge for prosecuting Kelvin Melton, but had accidentally grabbed her father.

Investigators determined that the Georgia phone sending the latest texts was also a burner, but this time, when agents approached cellular companies, their records proved more fruitful. The texts had been sent from Atlanta. Moreover, the phone had recently received a call from another number, which had been sending and receiving texts with yet another phone that police were able to determine was located inside the walls of Polk Correctional itself. That phone had placed almost a hundred calls to Melton's daughters.

The kidnapping, investigators came to believe, was being directed by Melton himself.

The FBI phoned Polk Correctional and told the warden to search Melton's cell. When Melton saw guards approaching, he barricaded the door and smashed his phone to pieces. It would take days to recover data from the device.

There was nothing the FBI could do to force Melton to cooperate with the investigators. He was already in prison for life. There was no additional information to be gleaned from any cellphone records. Agents had looked at surveillance tapes from the stores where the burner phones had been purchased and had scrutinized footage from cameras overlooking roads near Janssen's house. None of it was helpful. The FBI had hundreds of pieces of information. There were numerous dots, but nothing to connect them.

Some agents hoped the FBI's new computer system, a piece of complex software named Sentinel, might help unearth connections they had overlooked. Others were more skeptical. More than a decade earlier, the bureau had started building technologies that officials had promised would provide powerful new tools for solving crimes. Most of those efforts, however, were failures. One notable effort was abandoned in 2005 after $170 million was spent creating a search engine that crashed constantly. Another attempt was suspended in 2010 after auditors concluded it would cost millions more simply to figure out why the system wasn't working. A few years before Janssen was kidnapped, the agency's databases were still so outdated that most agents didn't even bother inputting the bulk of the information they collected during investigations. Instead, they used paper files and index cards, like their predecessors decades before.

Then, in 2012, the bureau had rolled out Sentinel. Simply put, it was a system for sorting and managing evidence, clues, witness testimony, and the tens of thousands of other little pieces of information agents collected every day. Sentinel was tied into analytical engines and databases that the bureau and other law-enforcement agencies had developed to look for patterns. The software's development had been overseen by a young man from Wall Street who had convinced the FBI to hire him by arguing that the bureau needed to draw on lessons from companies such as Toyota, and methods such as "lean manufacturing" and "agile programming." He had promised he could get Sentinel working in less than two years with a handful of software engineers—and then he had delivered.

Now Sentinel was functional. No one working on the Janssen case was certain if it would provide any help, but they were desperate. One of the agents began inputting each piece of information they had collected thus far, and then sat back to see if Sentinel would spit anything useful out.

II.

When Rick Madrid showed up for his job interview at the old General Motors plant, he wore mirrored shades, an Iron Maiden T-shirt, and a pair of cutoff jeans he had once described as "the greatest aphrodisiac in Northern California." It was 1984. Out of courtesy toward his interviewers—and because Madrid wanted this job—he had combed his beard and put on deodorant. He drew the line, however, at wearing sleeves that covered his tattoos.

Madrid was familiar with the plant in Fremont, California, because he had worked there until two years earlier, when GM had shut it down. Fremont was known, locally and nationally, as the worst auto factory in the world. Eight hours a day for twenty-seven years, Madrid had pounded rims into place with a sledgehammer, proselytized about the greatness of the United Auto Workers, and served rounds of "magic screwdrivers," a high-octane mixture of vodka and orange juice that he poured into plastic cups wedged into car frames so coworkers could partake as the vehicles progressed down the line. Fremont's assembly conveyors always moved smoothly, so the drinks hardly ever spilled. The bags of ice he put in the vehicles' trunks often warped the liners, but that was the problem of whoever bought the car. "Work was an interruption in my leisure time," Madrid later said. "I was there to earn money. I really didn't care about the quality of the job and neither did GM. They just wanted to get as many cars out as they could."

When Madrid showed up for this interview, however, he suspected things might be different this time. GM was partnering with the Japanese automaker Toyota to reopen the Fremont plant. For Toyota, this was a chance to build cars inside the United States and expand the company's sales in America. For General Motors, it was an opportunity to learn about the famed "Toyota Production System," which was producing cars of very high quality at very low costs in Japan. One hitch in the partnership was that GM's agreement with the UAW dictated that the plant had to hire at least 80 percent of its workers from employees who had been laid off two years earlier. So Madrid and his friends were showing up, one by one, to interview with New United Motor Manufacturing, Inc., or NUMMI.

Madrid figured he was a good candidate because his on-the-job drinking, truth be told, was tame compared to the antics of his former colleagues. Yes, he may have gotten drunk and had sex in the warehouse where they stored Chevy seats, but unlike many of his coworkers, he didn't snort coke while attaching brake pads or smoke weed from bongs built from muffler parts. He hadn't been a patron of the parking-lot RV where prostitutes offered services perfectly timed to union-mandated work breaks. Nor had Madrid ever deliberately sabotaged a vehicle like those who put empty whiskey bottles and loose screws behind door panels so they would bang around after the cars were sold.

The saboteurs were an extreme example of a fierce war that had consumed the Fremont plant in the GM days. Workers weren't above dirty tactics if they thought it would strengthen their union's hand. Employees knew that as long as they kept the assembly line moving, no one was likely to get punished for misbehavior, no matter how egregious. At GM, all that really mattered was keeping production on pace. Employees sometimes discovered mistakes on cars as they moved along the conveyor belt, but rather than stop and fix the problems, they would mark the vehicle with a wax crayon or a Post-it note and let it continue on its way. Eventually, those fully assembled autos would be hauled into the back lot and taken apart to repair the error. Once, a worker had a heart attack and fell into the pit as a car passed over; everyone waited for the vehicle to rumble along before they pulled him out. They all knew the plant's fundamental law: The line doesn't stop.

Madrid's first interview occurred in a small conference room. Across the table was a representative from the UAW, two Toyota executives from Japan, and a GM manager. Everyone exchanged pleasantries. They asked Madrid about his background and gave him some basic math and assembly problems to test his knowledge of auto manufacturing. They asked if he intended to drink while working. No, he said, he was done with that. It was a relatively brief conversation. Then, as he was walking out, one of the men from Japan asked Madrid what he had *disliked* about the plant when he worked there last time.

Rick Madrid had never been shy about speaking his mind. He didn't like working on cars he knew had problems, he told them, because whatever he was doing would have to be undone to repair a mistake. He didn't like that his suggestions were always ignored by his superiors. Once, he said, when a new tire mounting machine was being installed, he had come up with an idea for putting the controls in a different place to speed up work. He even sought out an engineer to show him a diagram of the concept. But when he came back from lunch, the new machine was in place with the controls in their original location. "I operated from the left side of the tire machine, all the controls were on the right side," he told the interviewers. "Thank goodness that engineer didn't build bridges."

When the plant was run by GM, workers were just cogs in a machine, Madrid told them. "You were just there to do what they told you to do," he said. Nobody ever asked him his opinion or cared what he thought.

He expressed all these frustrations to his interviewers and then kicked himself on the long drive home. He really needed this job. He should have kept his mouth shut.

A few days later, Madrid got the call. The Japanese executives had appreciated his honesty and were offering him a job. First, though, he would have to go to Japan for two weeks and learn about the Toyota Production System. Sixteen days later, NUMMI flew Madrid and about two dozen other workers to the Takaoka auto plant outside Toyota City, Japan, the first in a series of trips nearly every employee at NUMMI would take. When Madrid walked into the Japanese factory, he saw familiar assembly lines and heard the recognizable sounds of pneumatic tools hissing and buzzing. Why had they bothered flying him across the world to train inside a factory just like the one at home? After a basic tour and an orientation meeting, Madrid walked onto the factory floor and watched one man put bolts into doorframes, over and over, with an air-powered gun. By the time each car rolled off the line, Madrid knew, those bolts would be buried under layers of metal and plastic. It was just like California, except the signs were in Japanese and the bathrooms were much cleaner.

Then the worker manning the pneumatic gun pushed a bolt into place, applied his tool, and an ugly squeal sounded. The bolt had misthread the hole—a common mistake—and was stuck halfway in the doorframe. Madrid expected the man to signal the defect by marking the door, like they did at GM, so the car could be eventually towed to a back lot and repaired. The problem with that system, however, was that replacing the bolt would require disassembling the door, repairing the mistake, and then rebuilding everything. The trim would be less snug in the vehicle's frame afterward. Whoever bought the car wouldn't notice at first, but after a few years, the door would start jiggling. It would be a shoddier vehicle.

When the screw gun squealed inside the Japanese plant, though, something unexpected happened. The worker who made the mistake reached above his head and pulled a hanging cable that turned on a spinning yellow light. He then reversed the direction of his screw gun and pulled the bolt out of the doorframe, grabbed another tool, and used it to smooth the hole's threads. At this point, a manager walked over, stood behind the worker, and began asking questions. The worker ignored his boss except to bark out a few orders, and then grabbed another tool to rethread the hole. The conveyor belt was still moving, but the worker hadn't finished his repair. When the door got to the end of the worker's station, the entire assembly line stopped. Madrid had no idea what was going on.

Another man, clearly a senior manager, came over. Instead of yelling, he laid out a new bolt and equipment on a tray, like a nurse in an operating room. The worker kept issuing orders to his superiors. In Fremont, that would have gotten him slugged. Here, though, there were no angry shouts or anxious whispers. The other men on the line were calmly standing in place or double-checking parts they had just installed. No one seemed surprised at what was happening. Then the worker completed his rethreading, put a new bolt in the door, and pulled the cord above his head again. The assembly line started moving at normal speed. Everyone went back to work.

"I just didn't believe it," Madrid said. "Back home, I had seen a guy fall in the pit and they didn't stop the line. For so many years, I had learned you don't stop the line, no matter what." He had been told it cost $15,000 a minute to pause an assembly line. "But, for Toyota, quality came before income.

"That's when it dawned on me, that we can do this, we can compete against these guys by learning what they do," Madrid said. "One bolt, one bolt changed my attitude. I felt that I could finally, finally take pride in what I do."

As Madrid continued his training in Japan, there were other surprises as well. One day he shadowed a worker who, midway through a shift, told a manager he had an idea for a new tool that would help him install struts. The manager walked to the machine shop and returned fifteen minutes later with a prototype. The worker and manager refined the design throughout the day. The next morning, everyone had their own versions of the tool waiting at their stations.

Madrid's trainers explained that the Toyota Production System—which in the United States would become known as "lean manufacturing"—relied on pushing decision making to the lowest possible level. Workers on the assembly line were the ones who saw problems first. They were closest to the glitches that were inevitable in any manufacturing process. So it only made sense to give them the greatest authority in finding solutions.

"Every person in an organization has the right to be the company's top expert at something," John Shook, who trained Madrid as one of Toyota's first Western employees, told me. "If I'm attaching mufflers or I'm a receptionist or a janitor, I know more about exhaust systems or receiving people or cleaning offices than anyone else, and it's incredibly wasteful if a company can't take advantage of that knowledge. Toyota hates waste. The system was built to exploit everyone's expertise."

When Toyota had first proposed this management philosophy to General Motors, the Americans literally laughed at their naïveté. Maybe that approach works in Japan, they said, but it would fail in California. Workers at the Fremont plant didn't care about contributing expertise. What they cared about was doing as little work as possible.

"But the only way we would agree to the partnership was if GM promised to give this a try," said Shook. "Our basic philosophy was that no one goes to work wanting to suck. If you put people in a position to succeed, they will.

"What we didn't say was that if we couldn't figure out how to export the Toyota Production System, we were screwed," Shook said. "It's the *culture* that makes Toyota successful, not hanging cords or prototyping tools. If we couldn't export a culture of trust, we had no other ideas. So we sent everyone to America and prayed it would work."

● ● ●

In 1994, two business school professors at Stanford began studying how, exactly, one creates an atmosphere of trust within a company. For years, the professors—James Baron and Michael Hannan— had been teaching students that a firm's culture mattered as much as its strategy. The way a business treats workers, they said, was critical to its success. In particular, they argued that within most companies—no matter how great the product or loyal the customers—things would eventually fall apart unless employees trusted one another.

Then, each year, a few students would ask for evidence that supported those claims.

The truth was, Baron and Hannan *believed* their assertions were true, but they didn't have much data to back them up. Both men were trained as sociologists and could point to studies showing the importance of culture in making employees happy or recruiting new workers or encouraging a healthy work-life balance. But there were few papers showing how a company's culture impacted profitability. So in 1994, they embarked on a multiyear project to see if they could prove their assertion right.

First, though, they needed to find an industry that had lots of new companies they could track over time. It occurred to them that the flurry of technology start-ups appearing in Silicon Valley might provide the perfect sample. At the time, the Internet was in its infancy. Most Americans thought @ was something to ignore on a keyboard. Google was still just a number spelled "googol."

"We weren't inherently interested in tech and we had no idea the companies we were studying would become big deals," said Baron, who now teaches at Yale. "We just needed start-ups to study, and there were tech companies getting founded nearby and so every morning, we would buy the *San Jose Mercury News* and go through each page, and whenever a young company was mentioned, we would put our team into pursuit to find a phone number or mailing address, and then send someone to see if the CEO would answer a questionnaire." Over time, as they put it in a study they later wrote, "without realizing it when we started our study in 1994–1995, we assembled the most comprehensive database to date on the histories, structures, and HR practices of high-tech companies in Silicon Valley, just as the region was about to witness an economic and technological boom of historic proportions." The project ended up taking fifteen years and examining close to two hundred firms.

Their surveys looked at almost every variable that might influence a start-up's culture, including how employees were recruited, how applicants were interviewed, how much people were paid, and which workers executives decided to promote or fire. They watched college dropouts become billionaires and, in other cases, high-flying executives crash and burn.

Eventually, they collected enough data to conclude that most companies had cultures that fell into one of five categories. One was a culture they referred to as the "star" model. At these firms, executives hired from elite universities or other successful companies and gave employees huge amounts of autonomy. Offices had fancy cafeterias and lavish perks. Venture capitalists loved star model companies because giving money to the A-Team, conventional wisdom held, was always the safest bet.

The second category was the "engineering" model. Inside firms with engineering cultures, there weren't many individual stars, but engineers, as a group, held the most sway. An engineering mindset prevailed in solving problems or approaching hiring decisions. "This is your stereotypical Silicon Valley start-up, with a bunch of anonymous programmers drinking Mountain Dew at their computers," said Baron. "They're young and hungry and might be the next generation of stars once

STAR CULTURE

they prove themselves, but right now, they're focused on solving technical problems." Engineering-focused cultures are powerful because they allow firms to grow quickly. "Think of how fast Facebook expanded," said Baron. "When everyone comes from a similar background and mindset, you can rely on common social norms to keep everyone on the same path."

ENGINEERING CULTURE

The third and fourth categories of companies included those firms built around "bureaucracies" and those constructed as "autocracies." In the bureaucratic model, cultures emerged through thick ranks of middle managers. Executives wrote extensive job descriptions, organizational charts, and employee handbooks. Everything was spelled out, and there were rituals, such as weekly all-hands meetings, that regularly communicated the firm's values to its workers. An autocratic culture is similar, except that all the rules, job descriptions, and organizational charts ultimately point to the desires and goals of one person, usually the founder or CEO. "One autocratic chief executive told us that his cultural model was, 'You work. You do what I say. You get paid,'" Baron said.

BUREAUCRATIC   AUTOCRATIC
CULTURE   CULTURE

The final category was known as the "commitment" model, and it was a throwback to an age when people happily worked for one company their entire life. "Commitment CEOs say things like, 'I want to build the kind of company where people only leave when they retire or die,'" said Baron. "That doesn't necessarily mean the company is stodgy, but it does imply a set of values that might prioritize slow and steady growth." Some Silicon Valley executives told Baron they saw commitment firms as outdated, remnants of a corporate paternalism that had undermined industries such as American manufacturing. Commitment companies were more hesitant to lay people off. They often hired HR professionals when other start-ups were using precious dollars to recruit engineers or salespeople. "Commitment CEOs believe that getting the culture right is more important at first than designing the best product," Baron said.

COMMITMENT CULTURE

Over the next decade, Baron and Hannan kept close tabs on which start-ups thrived and which ones stumbled. About half the firms they studied remained in business for at least a decade; some became the most successful companies in the world. Baron and Hannan's goal was to see if particular corporate cultures were more likely to correlate with success. They were unprepared, however, for how dramatically the impact of culture came through. "Even in the fast-paced world of high-tech entrepreneurship in Silicon Valley, founders' employment models exert powerful and enduring effects on how their companies evolve and perform," the researchers wrote in 2002 in the journal *California Management Review.* The enormous impact of cultural decisions "is evident even after taking account of numerous other factors that might be expected to affect the success or failure of young technology ventures, such as company age, size, access to venture capital, changes in senior leadership, and the economic environment."

Just as Baron and Hannan had suspected, the star model produced some of the study's biggest winners. As it turned out, putting all the smartest people in the same room could yield vast influence and wealth. But, unexpectedly, star firms also failed in record numbers. As a group, they were less likely to make it to an IPO than any other category, and they were often beset by internal rivalries. As anyone who has ever worked in such a company knows, infighting is often more vicious inside a star-focused firm, because everyone wants to be *the* star.

In fact, when Baron and Hannan looked at their data, they found the only culture that was a consistent winner were the commitment firms. Hands down, a commitment culture outperformed every other type of management style in almost every meaningful way. "Not one of the commitment firms we studied failed," said Baron. "*None* of them, which is amazing in its own right. But they were also the fastest companies to go public, had the highest profitability ratios, and tended to be leaner, with fewer middle managers, because when you choose employees slowly, you have time to find people who excel at self-direction." Employees in commitment firms wasted less time on internal rivalries because everyone was committed to the company, rather than to personal agendas. Commitment companies tended to know their customers better than other kinds of firms, and as a result could detect shifts in the market faster. "Despite its being widely pronounced dead in Silicon Valley in the mid-1990s, the Commitment model fares very well in our sample," the researchers wrote.

"Venture capitalists love star firms because when you're investing in a portfolio of companies, all you need are a few huge successes," Baron told me. "But if you're an entrepreneur and you're betting on just one company, then the data says you're much better off with a commitment-focused culture."

One of the reasons commitment cultures were successful, it seemed, was because a sense of trust emerged among workers, managers, and customers that enticed everyone to work harder and stick together through the setbacks that are inevitable in any industry. Most commitment companies avoided layoffs unless there was no other alternative. They invested heavily in training. There were higher levels of teamwork and psychological safety. Commitment companies might not have had lavish cafeterias, but they offered generous maternity leaves, daycare programs, and work-from-home options. These initiatives were not immediately cost-effective, but commitment firms valued making employees happy over quick profits—and as a result, workers tended to turn down higher-paying jobs at rival firms. And customers stayed loyal because they had relationships that stretched over years. Commitment firms dodged one of the business world's biggest hidden costs: the profits that are lost when an employee takes clients or insights to a competitor.

"Good employees are always the hardest asset to find," said Baron. "When everyone wants to stick around, you've got a pretty strong advantage."

● ● ●

The first thing Rick Madrid did upon returning to California was tell everyone what he had seen in Japan. He talked about the hanging cables, called "andon cords," and about how managers took commands from workers instead of the other way around. He described watching assembly lines stop because some mechanic decided he needed more time to rebolt a door. He declared that everything at the Fremont plant was about to change now that NUMMI was in charge.

His friends were skeptical. They had heard this story before. GM had often said the company valued employee input—until employees began recommending changes that management didn't want to hear. In the weeks before the NUMMI plant opened, the factory's workers made sure their union memberships were up-to-date and held meetings to discuss tactics for fighting management, if it came to that. They voted to create a "NUMMI work-stoppage fund" to pay for workers' expenses if they went on strike. They demanded—and NUMMI immediately agreed to provide—a formal system for filing grievances.

Then, NUMMI's management announced the company's layoff policy. "New United Motor Manufacturing, Inc., recognizes that job security is essential to an employee's well-being," read the company's agreement with the United Auto Workers. "The Company agrees that it will not lay off employees unless compelled to do so by severe economic conditions that threaten the long-term viability of the Company." NUMMI promised it would cut executive pay rather than fire workers and train people to sweep floors, repair machines, or serve meals in the cafeteria to preserve their jobs. Every employee complaint and suggestion, no matter how far-fetched or expensive, would be implemented, or a response would be publicly posted explaining why not. Every team was given authority to change their stations' layouts and work flow. Anyone, at any time, could stop the assembly line if they saw a problem. No American car company had ever made so public a promise to avoid layoffs and respond to worker complaints.

Skeptical workers said that such pledges were easy to make when the plant wasn't even operating yet, but they grudgingly agreed to play along. The factory began producing Chevy Novas on December 10, 1984.

Rick Madrid was assigned to a team that stamped out hoods and doors from giant sheets of steel. It was immediately clear to him that things were different. People who once sought trysts in the storage room kept their hands to themselves. There was no obvious drinking on the job. The RV hadn't returned to the parking lot. People were scared to try anything. They didn't want to push their luck. This hesitancy, however, also had less useful consequences. No one was pulling andon cords or making suggestions because no one was eager to cost the factory $15,000 a minute. No one was sure it wouldn't cost them their job.

A month after the plant reopened, Tetsuro Toyoda—NUMMI's president, whose grandfather had founded Toyota in 1933—walked the Fremont floor. He saw an employee struggling to install a taillight that was wedged into a car frame at an odd angle. Toyoda approached the worker and, reading the name stitched on his uniform, said, "Joe, please pull the cord."

"I can fix this, sir," Joe said.

"Joe, please pull the cord."

Joe had never pulled an andon cord. No one in his area had. Since the plant had opened, there had been only a few cord pulls, and one of those had been an accident.

"Sir, I can fix this," Joe said, working furiously to pop the taillight into place.

Joe's team leader was standing nearby. That man's manager had been shadowing Toyoda as he walked through the factory, so he was orbiting nearby as well. When Joe glanced up, he saw a half dozen of the plant's most senior executives staring back.

"Joe, please," Toyoda said. Then he stepped over, took Joe's hand in his own and guided it to the andon cord, and together they pulled. A flashing light began spinning. When the chassis reached the end of Joe's station without the taillight correctly in place, the line stopped moving. Joe was shaking so much, he had to hold his crowbar with both hands. He finally got the taillight positioned and, with a terrified glance at his bosses, reached up and pulled the andon cord, restarting the line.

Toyoda faced Joe and bowed. He began speaking in Japanese.

"Joe, please forgive me," a lieutenant translated. "I have done a poor job of instructing your managers of the importance of helping you pull the cord when there is a problem. You are the most important part of this plant. Only you can make every car great. I promise I will do everything in my power to never fail you again."

By lunchtime, everyone inside the factory had heard the story. The next day, andon cords were pulled more than a dozen times. The next week, more than two dozen times. A month later, the plant was averaging nearly a hundred pulls a day.

The importance of the andon cords and the employee suggestions and Toyoda's apology was that they demonstrated that the fate of the company was in the employees' hands. "There was a genuine devotion to convincing employees they were part of a family," said Joel Smith, the UAW representative at NUMMI. "It had to be reinforced constantly, but it was real. We might have disagreements or see things differently, but at the end of the day, we were committed to each other's success."

"If people started pulling andons for no good reason, the plant would have fallen apart," said Smith. Everyone knew it still cost thousands of dollars each minute a line was stopped, "and that anyone could stop the line, at any time, without penalty. So employees could bankrupt the place if they wanted to.

"Once you're entrusted with that kind of authority, you can't help feel a sense of responsibility," said Smith. "The most junior workers didn't want NUMMI to go bankrupt, and the management didn't want that, and so, suddenly, everyone was on the same side of the table." And as workers were empowered to make more choices, their motivation skyrocketed. Just as Mauricio Delgado and the U.S. Marine Corps had found in other settings, when workers felt a greater sense of control, their drive expanded.

Word of the NUMMI experiment spread quickly. When professors from Harvard Business School visited a few years after the plant reopened, they found that former GM employees who once spent only forty-five seconds of every minute working now averaged fifty-seven seconds of labor per minute. By 1986, "NUMMI's productivity was higher than that of any other GM facility and more than twice that of its predecessor, GM-Fremont," they wrote. Absenteeism had dropped from 25 percent during the GM days to 3 percent under NUMMI. There were no observable levels of substance abuse, prostitution, or sabotage. The formal grievance system was hardly ever used. NUMMI's productivity was as high as that of plants in Japan, "even though its workers were, on average, ten years older and much less experienced with the Toyota production system," the Harvard researchers wrote. In 1985, *Car and Driver* magazine printed an issue with the cover line "Hell Freezes Over," announcing NUMMI's accomplishments. The worst auto factory on earth had become one of the most productive plants in existence, using the same workers as before.

Then, four years after NUMMI opened, the recession hit the auto industry. The stock market crashed. Unemployment was rising. Car sales plummeted. NUMMI's managers estimated they needed to reduce production by 40 percent. "Everyone was saying there were going to be layoffs," said Smith, the UAW rep. Instead, the plant's top sixty-five executives all took pay cuts. Assembly line workers were reassigned to janitorial duties or landscaping, or sent into the paint room to scrape air vents instead of let go. The company proved it was committed.

"After that, workers were willing to do anything for the company," Smith said. "Four separate sales slumps over thirty years, and NUMMI never did layoffs once. And each time, when the business finally came back, everyone worked harder than before."

Rick Madrid retired from NUMMI in 1992, after almost four decades of building cars. Three years later, the Smithsonian mounted an exhibit at the National Museum of American History that included Madrid's ID badge and his hat in a show named *A Palace of Progress*. NUMMI, the curators wrote, was iconic, a factory that had demonstrated it was possible to unite workers and managers around a common cause through mutual commitment and shared power.

Even now NUMMI is cited inside business schools and by corporate chieftains as an example of what organizations can achieve when a commitment culture takes hold. Since NUMMI was founded, the "lean manufacturing" principles have infiltrated nearly every corner of American commerce, from Silicon Valley to Hollywood to healthcare. "I'm really glad I ended up my years as an autoworker at NUMMI," Madrid said. "I went from being depressed, bored, people didn't even know I existed, to seeing J. D. Power name NUMMI as a top quality plant."

NUMMI's workers got together for a party after that J. D. Power announcement. "And when I spoke, I said, we're the best damn autoworkers in the world," said Madrid. "Not just the workers. Not just the managers. All of us, together, are the best, because we're devoted to each other."

III.

Six years before Frank Janssen was kidnapped, the Federal Bureau of Investigation had reached out to a thirty-four-year-old Wall Street executive to see if he would be interested in overseeing development of the bureau's technology systems. Chad Fulgham had never worked in law enforcement before. His specialty was developing large computer networks for investment banks such as Lehman Brothers and JPMorgan Chase. So he was surprised when he received a call in 2008 explaining that the FBI wanted an interview.

Improving the bureau's technology had long been a priority for federal officials. As early as 1997, the FBI's top leaders had promised Congress they would deliver an overhauled system that tied together the dozens of internal databases and analytical engines the bureau maintained. This network, officials said, would give agents powerful new tools for connecting the dots among disparate cases. But by the time the bureau contacted Fulgham eleven years later, work on that system, Sentinel, had already consumed $305 million with no end in sight. The agency had brought in an outside group to figure out why Sentinel was taking so long. The experts said the bureau was so bogged down by bureaucracy and conflicting agendas that it would take tens of millions of dollars simply to get the program back on track.

The bureau called Fulgham to see if he could find a cheaper way to get things moving again. "I secretly always wanted to work for the FBI or the CIA," he told me. "So when they called me, particularly with this huge, hairy problem, it was like getting offered my dream job."

First, though, Fulgham needed to convince the bureau that his approach was the right one. Fulgham's management style, he explained, drew its inspiration from examples like NUMMI. In the previous two decades, as NUMMI's success had become better known, executives in other industries had started adapting the Toyota Production System philosophy to other industries. In 2001, a group of computer programmers had gathered at a ski lodge in Utah to write a set of principles, called the "Manifesto for Agile Software Development," that adapted Toyota's methods and lean manufacturing to how software was created. The Agile methodology, as it came to be known, emphasized collaboration, frequent testing, rapid iteration, and pushing decision making to whoever was closest to a problem. It quickly revolutionized software development and now is the standard methodology among many tech firms.

Among filmmakers, the "Pixar method" was modeled specifically on Toyota's management techniques and became famous for empowering low-level animators to make critical choices. When Pixar's leadership was asked to take over Disney Animation in 2008, executives introduced themselves with what became known as "the Toyota Speech," "in which I described the car company's commitment to empowering its employees and letting people on the assembly line make decisions when they encountered problems," Pixar cofounder Ed Catmull later wrote. "I stressed that no one at Disney needed to wait for permission to come up with solutions. What is the point of hiring smart people, we asked, if you don't empower them to fix what's broken?"

In hospitals, the distribution of authority to nurses and others who are not physicians is referred to as "lean healthcare." It is a management philosophy and a "culture in which anyone can, and indeed must, 'stop the line,' or stop the care process if they feel something is not right," the chairman of one lean hospital, the Virginia Mason Medical Center, wrote in 2005.

These approaches emerged in different industries, but they and other adaptations of lean manufacturing shared key attributes. Each was dedicated to devolving decision making to the person closest to a problem. They all encouraged collaboration by allowing teams to self-manage and self-organize. They emphatically insisted on a culture of commitment and trust.

Fulgham argued that the FBI's technology efforts could succeed only if the bureau embraced a similar approach. FBI officials had to commit to distributing critical decision-making power to people on the ground, he said, such as lowly software engineers or junior field agents. This approach was a significant shift because previously, bureau executives—distrustful of one another and consumed by internal power struggles—had designed new technology systems by first outlining thousands of specifications each piece of software needed to satisfy. Committees filled hundreds of pages with rules for how databases ought to function. Any major change required approvals from numerous officials. The system was so dysfunctional that software development teams would sometimes spend months building a program, only to be told it was canceled when they were done. And the results were often dysfunctional as well. When Fulgham asked for a demonstration of the Sentinel work done thus far, for instance, an engineer led him to a computer monitor and invited him to input some keywords, such as a criminal's alias and an address associated with a crime.

"In fifteen minutes, we'll have a report of previous cases linked to that address and name," the engineer said.

"The people I'm going to report to carry guns, and you want me to tell them it will take the computer fifteen minutes to provide help?" said Fulgham.

A 2010 inspector general's report had said it would take another six years and $396 million to get Sentinel working. Fulgham told the bureau's director that if they gave him the authority to distribute control, he would cut the number of people needed from more than four hundred to just thirty employees and deliver Sentinel for $20 million in a bit over a year. Soon Fulgham and a team of software engineers and FBI agents were holed up in the basement of the bureau's Washington, D.C., headquarters. The only rules, Fulgham told them, were that everyone had to make suggestions, anyone could declare a time-out if they thought a project was moving in the wrong direction, and the person closest to a problem had primary responsibility for figuring out how to solve it.

The main problem with Sentinel, Fulgham believed, was that the bureau—like many big organizations—had tried to plan everything in advance. But creating great software requires flexibility. Problems pop up unexpectedly and breakthroughs are unpredictable. The truth was, no one knew exactly how FBI agents would use Sentinel once it was functional, or how it would need to change as crime-fighting techniques evolved. So instead of meticulously predesigning each interface and system—instead of trying to control from above—they needed to make Sentinel into a tool that could adapt to agents' needs. And the only way to do that, Fulgham was convinced, was if developers were unfettered themselves.

Fulgham's team started by coming up with more than one thousand scenarios in which Sentinel could be useful, everything from inputting victims' statements to tracking evidence to interfacing with FBI databases that looked for patterns among clues. Then they started working backward to figure out what kind of software should accommodate each need. Every morning, the team conducted a "stand-up"—meetings where everyone stood to encourage brevity—and recounted the previous day's work and what they hoped to accomplish over the next twenty-four hours. Whoever was closest to a particular problem or a piece of code was considered the expert on that topic, but any programmer or agent, no matter their rank, was free to make suggestions. In one case, a programmer and a field agent, after brainstorming, suggested that they model part of Sentinel on TurboTax, the popular financial software that reduced thousands of pages of complicated tax laws into a series of basic questions. "The idea was basically 'Investigations and Justice for Dummies,'" said Fulgham. "It was absolutely brilliant."

Under the old system, getting approval for that proposal would have taken upward of six months and required dozens of memos, each carefully scrubbed of any mention of TurboTax or any indications that programmers intended to simplify federal procedures. No one would have wanted an enterprising lawyer or journalist getting their hands on something that used plain English to explain how the system worked. Under Fulgham, though, none of that bureaucracy existed. The programmer and agent mentioned the idea on a Monday, had a prototype ready by Wednesday, and everyone agreed to use the approach going forward on Friday. "It was like government on steroids," Fulgham said.

Every two weeks, the team demonstrated their work for a broad audience of high-ranking officials who provided feedback. The bureau's director had forbidden anyone from micromanaging or making demands. At most, division heads could offer suggestions, each of which was cataloged and evaluated by whoever was closest to that piece of code. Gradually, the Sentinel team became bolder and more ambitious, not just building systems for record keeping, but linking Sentinel to tools that identified trends and threats and made comparisons among cases. By the time they were done, Sentinel was at the core of a system so powerful it could look across millions of investigations simultaneously and pick out patterns agents had missed. The software went live sixteen months after Fulgham took over. "Deployment of the Sentinel application in July 2012 represented a pivotal moment for the FBI," the agency later wrote. Sentinel, in its first month alone, was used by more than thirty thousand agents. Since then, it has been credited with helping solve thousands of crimes.

At NUMMI, decentralizing decision making helped inspire a workforce. At the FBI, it played a different role. Lean management and agile methods helped fuel the ambitions and innovation of junior programmers who had been beaten down by bureaucracy. It emboldened them to come up with solutions no one had considered before. It convinced people to swing for the fences because they knew they wouldn't be punished if they missed the ball now and then.

"The effect of Sentinel on the FBI has been dramatic," wrote Jeff Sutherland, one of the authors of the Agile Manifesto, in a study of Sentinel's development published in 2014. "The ability to communicate and share information has fundamentally changed what the bureau is capable of."

More important, the *way* that Sentinel succeeded served as a source of inspiration to the agency and its leadership. "The Sentinel experience taught us a lot about how much potential can be unlocked when you give people more authority," Jeff Johnson, the bureau's current chief technology officer, told me. "We saw how much more passionate people can become. You look at some of our recent cases—the kidnapping in North Carolina, hostage rescue situations, terrorism investigations—and in situations like that, we've learned it's critical that agents feel like they can make independent decisions.

"But empowering people in an agency this size is really hard," Johnson said. "That was one of the problems before 9/11—people didn't feel rewarded for independent thinking. Then you look at things like the development of Sentinel, and you see how much is possible."

IV.

After the agents working on the Frank Janssen kidnapping case inputted the data they had collected into Sentinel, the software and the databases connected to it began looking for patterns and leads. The agents had entered the cellphone numbers the bureau had collected, the addresses investigators had visited, and the aliases the kidnappers had used in intercepted calls. Others inputted the names of people who had visited Kelvin Melton in prison, license plates caught on cameras around Janssen's home, and credit card transactions from inside the stores where the burner phones were bought. Every detail was fed into Sentinel in the hopes a connection would emerge.

Eventually, the agency's databases discovered a coincidence: The phone that had sent photos of Frank Janssen to his wife had also made a call to Austell, Georgia, a small city outside Atlanta. The FBI's computers had looked through millions of records from other cases, and had found a link to Austell from another case.

In March 2013, a year earlier, a confidential informant had given the bureau the address of an apartment in Austell that he said criminals used as a safe house. That same informant, in a different conversation, had also mentioned an imprisoned gang leader who had "put a hit on the female District Attorney who prosecuted him," a reference the FBI believed was to Kelvin Melton, the man who allegedly planned the Janssen kidnapping.

At the time of those conversations, no one inside the FBI knew what the informant was talking about. Janssen wouldn't be kidnapped until a year later. And since then, no one had given the conversation a second thought. The agents who had interviewed the informant weren't even part of the team looking for Janssen.

But now, the computers connected to Sentinel found a link: A confidential informant had mentioned someone who fit the description of Kelvin Melton, who had allegedly planned the kidnapping. That informant had also mentioned an apartment in Austell—an apartment that, Sentinel had just revealed, might have received a phone call from one of the kidnappers' phones.

Someone needed to visit that apartment.

The problem was, this was just one of dozens of leads investigators were chasing. There were former associates of Melton's to track down, prison visitors to scrutinize, former girlfriends who might be involved. There were too many potential leads, in fact, for agents to pursue them all. The FBI needed to prioritize, and it wasn't clear that chasing a clue from a year-old conversation was the best use of time.

In recent years, however, as the success of Sentinel had attracted more notice within the bureau, officials had become increasingly committed to using lean and agile techniques throughout the agency. Commanders and field agents had embraced the philosophy that the person closest to a question should be empowered to answer it. FBI director Robert Mueller had launched a series of initiatives— the Strategy Management System, the Leadership Development Program, Strategic Execution Teams—that were designed to spark, as he told Congress in 2013, "a paradigm shift in the FBI's cultural mindset." One particular focus was encouraging junior agents to make independent decisions about which leads they should pursue, rather than waiting for orders from superiors. Any agent could chase a clue if they thought something was being overlooked. It was a law enforcement version of pulling the andon cord. "It's a critical shift," said Johnson, the FBI chief technology officer. "The people closest to the investigation *have* to be empowered to make choices about how they spend their time." Sentinel wasn't the only influence behind this change, but it accelerated the adoption of an agile philosophy inside the bureau. "The FBI's basic mindset is agile now," Fulgham told me. "Sentinel's success solidified that."

The investigators on the Janssen case had dozens of leads to choose from. But junior agents were encouraged to make decisions for themselves. So two young investigators decided to visit the apartment the confidential informant had mentioned over a year ago.

When they arrived at the apartment, they learned it was occupied by a woman named Tianna Brooks. She wasn't home, but her two young children were there, unsupervised. The agents called child protective services, and once the kids had been collected by social workers, the agents began canvassing neighbors, asking where Brooks had gone. No one knew, but one person said Brooks had been visited by two men staying nearby. The agents found those men and questioned them. They said they didn't know anything about Brooks or any kidnappings.

At 11:33 P.M., a call came in to one of the many phones the FBI had linked to the kidnappers and, as a result, were under surveillance.

"They got my kids!" a woman's voice said.

The agents in Austell were told about the call, and began questioning their suspects more forcefully. The agents pointed out that the two suspects had recently visited Tianna Brooks. Now, the FBI had intercepted a telephone call of a panicked woman—possibly Brooks herself—saying the FBI had her children.

In other words, the two suspects had recently visited someone who may be linked to a kidnapping.

Was there anything else they wanted to say?

One of them mentioned an apartment in Atlanta.

The agents radioed their colleagues at the kidnapping command center and a few minutes before midnight, SWAT trucks arrived at the Atlanta apartment complex the suspects had mentioned. Officers jumped out of the vehicles and ran past dilapidated buildings. They paused in front of one home and rammed their way through a wrought-iron door. Inside were two men sitting in chairs with guns next to them, caught completely off guard. The room also contained ropes, a shovel, and bottles of bleach. The men had recently used their phones to send texts about how to dispose of a body. "Get bleach and throw it on the walls," someone had ordered them. "Maybe do it in the closet."

An officer in riot gear ran into a bedroom and tore open all the doors. Inside the closet, he found Frank Janssen tied to a chair, unconscious, with blood still on his face from where the assailants had pistol-whipped him. He had been missing for six days by then, and was severely dehydrated. The police cut him free and carried him out of the apartment, past where Janssen's attackers lay on the floor, hands cuffed behind their backs. Janssen was put into an ambulance and rushed to the hospital. When his wife saw him, she began sobbing. For almost a week, no one had known if he was alive or dead. And now here he was, with no serious injuries beyond bruises and cuts. He was released two days later with a clean bill of health.

The breakthrough in the Janssen case didn't occur simply because the bureau's computer systems connected the dots between his kidnapping and an old, seemingly unrelated interview with a confidential informant. Rather, Janssen was rescued because hundreds of dedicated people worked nonstop to chase dozens of leads, and because an agile culture empowered junior agents to make independent decisions and follow the clues *they* thought made sense.

"Agents learn to investigate by listening to their guts and learning they can change direction when new evidence appears," Fulgham told me. "But for those instincts to be unlocked, management has to empower them. There has to be a system in place that makes you trust that you can choose the solution you think is best and that your bosses are committed to supporting you if you take a chance that might not pay off. That's why agile has been embraced at the bureau. It talks to who they are."

This, ultimately, is one of the most important lessons of places such as NUMMI and the lean and agile philosophies: Employees work smarter and better when they believe they have more decision-making authority *and* when they believe their colleagues are committed to their success. A sense of control can fuel motivation, but for that drive to produce insights and innovations, people need to know their suggestions won't be ignored, that their mistakes won't be held against them. And they need to know that everyone else has their back.

The decentralization of decision making can make anyone into an expert—but if trust doesn't exist, if employees at NUMMI don't believe management is committed to them, if programmers at the FBI aren't trusted to solve problems, if agents aren't encouraged to follow a hunch without fear of admonishment, organizations lose access to the vast expertise we all carry within our heads. When people are allowed to stop the assembly line, redirect a huge software project, or follow an instinct, they take responsibility for making sure an enterprise will succeed.

A culture of commitment and trust isn't a magic bullet. It doesn't guarantee that a product will sell or an idea will bear fruit. But it's the best bet for making sure the right conditions are in place when a great idea comes along.

That said, there are good reasons companies don't decentralize authority. There is a powerful logic behind investing power in only a few hands. At NUMMI, a small group of disgruntled workers could have bankrupted the firm by pulling andon cords needlessly. Inside the FBI, a misguided programmer could have built the wrong computer system. An agent might have followed the wrong hunch. But, in the end, the rewards of autonomy and commitment cultures outweigh the costs. The bigger misstep is when there is never an opportunity for an employee to make a mistake.

A few weeks after his rescue, Frank Janssen sent a thank-you letter to the agents who rescued him. "I have never felt a greater feeling of joy, relief, and freedom than that miraculous moment when I heard a firm American soldier's voice say, 'Mr. Janssen, we are here to take you home,'" he wrote. "Despite the nightmare that I experienced, the fact that I am writing this letter from the comfort of my home is a testament to the many wonderful things that were done by many wonderful people." It was a calamity that he was kidnapped, Janssen wrote, and a testament to the commitment of the FBI that he was saved.

6

DECISION MAKING

Forecasting the Future (and Winning at Poker) with Bayesian Psychology

The dealer looks at Annie Duke and waits for her to say something. There is a pile of chips worth $450,000 in the middle of the table and nine of the world's best poker players—all men, except for Annie—impatiently waiting for her to bet. It's the 2004 Tournament of Champions, a televised competition with $2 million to the winner. There is no prize for second place.

The dealer hasn't put down any communal cards yet, and Annie is holding a pair of tens. Her hand is strong—strong enough that she has already shoved most of her chips into the pot. Now she has to decide if she wants to bet everything. All the other players have folded except for one—Greg Raymer, aka "the FossilMan," a rotund gentleman from Connecticut who carries pieces of petrified bark in his pockets and wears sunglasses with holographic lizard eyes.

Annie doesn't know what cards the FossilMan is holding. Until a few seconds ago, based on how things were proceeding, she figured she was going to win this hand. But then the FossilMan pushed everything onto the pot and threw a wrench in Annie's plans. Has the FossilMan been playing her this whole time? Luring her into bigger and bigger bets while waiting to pounce? Or is he trying to scare her off with a wager so large he thinks she'll get spooked and walk away?

Everyone is staring at Annie. She has no idea what to do.

She could fold. But that would mean forfeiting the tens of thousands of dollars she's spent to get to this table, all the progress she's made over the past nine hours, everything she's worked so hard to earn.

Or she could match his wager and bet everything. If she loses, she'll be knocked out of the tournament. If it pays off, though, and she wins this hand, she'll instantly become the tournament's frontrunner, a step closer to paying for her kids' school bills and her mortgage, not to mention her messy divorce and all the uncertainties that give her stomachaches at night.

She looks again at the mountain of chips on the table and feels a pressure rising in her throat. She's had panic attacks all her life, breakdowns so severe that she used to lock herself inside her apartment and refuse to leave. Twenty years ago, during her sophomore year at Columbia University, she became so anxious she walked into a hospital, begged them to admit her, and didn't come out for two weeks.

Forty-five seconds pass while Annie tries to figure out what to do. "I'm so sorry," she says. "I know I'm taking too long. This is just a really hard decision."

Annie focuses on her pair of tens. She thinks about what she knows and doesn't know. What Annie likes about poker are the certainties. The trick to this game is making predictions, imagining alternative futures and then calculating which ones are most likely to come true. Statistics make Annie feel in control. She might not know exactly what's coming, but she knows the precise likelihoods of being right or wrong. The poker table feels serene.

And now the FossilMan has blown that peacefulness to hell by making a bet that doesn't match any of the scenarios inside Annie's head. She has no idea how to gauge what is most likely to occur. She's frozen.

"I'm really sorry," she says. "I just need a second more."

● ● ●

Many afternoons during Annie's childhood, her mother would sit at the kitchen table with a pack of cigarettes, a glass of scotch, and a deck of cards, and play hand after hand of solitaire until the alcohol was gone and the ashtray was full. Then she would stagger to the couch and fall asleep.

Annie's father was an English teacher at St. Paul's School in New Hampshire, a boarding school for the scions of senators and CEOs. Her family lived in a house attached to one of the dorms and so whenever her parents fought about her mom's drinking or her father's lack of money—which they did frequently—Annie was sure her classmates could overhear. She often felt like an outcast at the school, too poor to vacation with the rich kids, too smart to bond with the popular girls, too anxious to be comfortable among the hippies, too interested in math and science for student government. For Annie, the key to surviving amid the shifting tectonics of teenage popularity was learning to forecast. If she could predict which students' social capital was rising or falling, it was easier to avoid the infighting. If she could predict when her parents were arguing or her mom was drinking, she would know if it was safe to bring classmates home.

"When you have an alcoholic parent, you spend a lot of time thinking about what's coming," Annie told me. "You never take for granted that you'll get dinner or that someone will tell you when to go to bed. You're always waiting for everything to fall apart."

After graduation, Annie left for college at Columbia and soon discovered the psychology department. Here, at last, was what she had been looking for. There were classes that reduced human be-havior to understandable rules and social formulas; teachers who gave lectures on the different categories of personality and why anx-ieties emerge; studies about the impact of having an alcoholic par-ent. She felt like she was starting to understand why she sometimes had panic attacks, why it occasionally felt impossible for her to leave her bed, why she carried this dread that something bad might hap-pen at any time.

Psychology, at that moment, was undergoing a transformation brought on by discoveries in cognitive sciences that were bringing a scientific rigor to understanding behaviors that had long seemed immune to methodical analysis. Psychologists and economists were working together to understand the codes that explain why people do what they do. Some of the most exciting research—work that would eventually win a Nobel Prize—was focused on studying how people make decisions. Why, researchers wondered, do some peo-ple decide to have children when the costs, in terms of money and hard work, are so obvious, and the payoffs, such as love and content-ment, are so hard to calculate? How do people decide to send their kids to expensive private schools instead of free public ones? Why does someone decide to get married after playing the field for years?

Many of our most important decisions are, in fact, attempts to forecast the future. When we send a child to private school, it is, in part, a bet that money spent today on schooling will yield happiness and opportunities in the future. When we decide to have a baby, we're forecasting that the joy of becoming a parent will outweigh the cost of sleepless nights. When we choose to get married—though it may seem completely unromantic—we are, at some level, calculat-ing that the benefits of settling down are greater than the oppor-tunity of seeing who else comes along. Good decision making is contingent on a basic ability to envision what happens *next*.

What fascinated psychologists and economists was how fre-quently people managed, in the course of their everyday lives, to choose among various futures without becoming paralyzed by the complexities of each choice. What's more, it appeared that some people were more skilled than others at envisioning various futures and choosing the best ones for themselves. Why were some people able to make better decisions?

When Annie graduated from college, she enrolled in a PhD program in cognitive psychology at the University of Pennsylvania and began collecting grants and publications. After five years of hard work and a successful run of papers and awards, with only months to go to her doctorate, she was invited to give a series of "job talks" at several universities. If she performed well, she was practically guaranteed a prestigious professorship.

The night before her first speech at New York University, she took the train to Manhattan. She had been feeling anxious all week. At dinner she began throwing up. She waited an hour, drank a glass of water, and threw up again. She couldn't turn off her anxiety. She couldn't stop thinking that she was making a mistake, that she didn't want to be a professor, that she was only doing all this because it had seemed like the safest, most predictable path. She called NYU to postpone her talk. Her fiancé flew to Manhattan and took her back to Philadelphia, where she checked into a hospital. She was discharged weeks later, but even then her anxiety was like a hot stone in her stomach. She went straight from the hospital to a classroom at Penn where she was supposed to teach and somehow made it through the lecture, so nauseous and jumpy she almost fainted. She couldn't teach another class, she decided. She couldn't give her job talks. She couldn't become a professor.

She shoved her research into the trunk of her car, sent a note to her professors saying that she would be hard to reach for a while, and drove west. Her fiancé had found a house that cost $11,000 outside Billings, Montana. When Annie arrived, she determined that, even at that price, they had paid too much. But by then, she was too exhausted to do anything about it. She put her dissertation materials into the closet and settled onto the couch. Her only goal was to think as little as possible.

A few weeks later, her brother, Howard Lederer, called to invite her to Las Vegas for a vacation. Howard was a professional poker player, and every spring for the past few years, he had flown Annie out to sit by the swimming pool of the Golden Nugget while he played in a tournament. Whenever she got bored, she would wander inside to watch him compete or play a few hands of poker herself. When he called this year, however, Annie said she was too sick to make the trip.

Howard was concerned. Annie loved Vegas. She never turned down a trip.

"Why don't you at least find a local poker game?" he said. "It might help you get out of the house."

By then she was married, and so she asked her husband to make some inquiries. They learned there was a bar in Billings named the Crystal Lounge where retired ranchers, construction workers, and insurance agents played poker every afternoon in the basement. It was a smoke-filled, joyless dungeon. Annie went one afternoon and loved it. She went again a few days later and walked out fifty dollars richer. "Playing poker down there was this combination of math, which I loved, and all of the cognitive science stuff I had been doing in graduate school," Annie told me. "You could watch people try to bluff each other and hide their excitement when they got a good hand, and all these other kinds of behaviors we'd spent hours talking about in classes. Every night, I would call my brother and talk through the hands I had played that day, and he would explain my mistakes, or how someone else had figured out my game and had started using that against me or what I should do different next time." Initially, she wasn't very good. But she won often enough to keep going. She noticed that her stomach never hurt at the poker table.

Pretty soon, she was going to the Crystal Lounge every weekday, like a job, arriving at three P.M. and staying until midnight, taking notes and testing strategies. Her brother sent her a check for $2,400 with the agreement that he would get half her winnings. She was up $2,650 by the end of the first month even after his cut. The next spring, when he invited her to Vegas, she drove fourteen hours, bought a seat in a tournament, and by the end of the first day had $30,000 in chips.

Thirty thousand dollars was more than Annie had ever earned in a full year as a grad student. She understood poker—understood it better than many of the people she was playing against. She understood that a losing hand isn't necessarily a loss. Rather, it's an experiment. "The thing I had figured out by that point was the difference between intermediate and elite players," Annie told me. "At the intermediate level, you want to know as many rules as possible. Intermediate players crave certainty. But elite players can use that craving against them, because it makes intermediate players more predictable.

"To be elite, you have to start thinking about bets as ways of asking other players questions. Are you willing to fold right now? Do you want to raise? How far can I push before you start acting impulsively? And when you get an answer, that allows you to predict the future a little bit more accurately than the other guy. Poker is about using your chips to gather information faster than everyone else."

By the end of the tournament's second day, Annie had $95,000 in chips. She finished in twenty-sixth place, ahead of hundreds of professionals, some with decades of experience. Three months later, she and her husband moved to Las Vegas. At some point, she called her professors at Penn and told them she wasn't coming back.

● ● ●

A full minute has passed. Annie still has a pair of tens. If the Fossil-Man is holding a higher pair—say, two queens—and Annie stays in the hand, it's almost certain she'll be eliminated from the Tournament of Champions. But if she wins the hand, she'll become the table's chip leader.

All of the odds and probability charts floating around Annie's head are telling her to do one thing: Match the FossilMan's bet. But every time Annie has asked the FossilMan a question during this tournament by placing a wager, he's answered with a highly rational response. He's never put everything on the line without a good reason. Now, in this hand, he's pushed all of his chips into the pot, even as Annie has raised again and again.

Annie is aware that the FossilMan knows how hard it is for her to back down at this point. He knows that, unlike some of the other people at this table, she isn't in the Poker Hall of Fame. This is her first time in front of a million television viewers. He might even know that she's worried she doesn't belong here at all, that she suspects she was only invited because the TV producers wanted a woman at the table.

Annie suddenly realizes she's been thinking about this hand wrong all along. The FossilMan has been betting as if he has a good hand because, in fact, he has a good hand. Annie has been overthinking—or, at least, she thinks she's been overthinking. She's not sure.

She looks at her pair of tens, looks at the $450,000 on the table, and folds. The FossilMan takes the money. Annie has no idea if she just made a good or bad choice because the FossilMan doesn't have to show anyone his cards. Another player leans over. You completely misread the situation, he whispers to her. If you had stayed in, you would have won.

A few hands later, Annie has already folded when the FossilMan, with a ten and a nine, bets all his chips once again. It's a smart play, the right move, but as the other cards fall on the table, they go against him. Even the smartest poker players can be undone by bad luck. Probabilities can help you forecast likelihoods, but they can't guarantee the future. Just like that, the FossilMan is out of the tournament. As he stands to leave, he leans over to Annie.

"I know the hand you had earlier was really hard for you," he tells her. "I want you to know that I had two kings and you made a good fold."

When he says that, the knot of panic in her stomach melts. She's been distracted ever since she folded against him. She's been second-guessing herself, turning the hand over in her mind, trying to figure out if she played it right or wrong. Now her head is back in the game.

It's normal, of course, to want to know how things will turn out. It's scary when we realize how much rides on choices where we can't predict the future. Will my baby be born healthy or sick? Will my fiancée and I still love each other ten years from now? Does my kid need private school or will the local public school teach her just as much? Making good decisions relies on forecasting the future, but forecasting is an imprecise, often terrifying, science because it forces us to confront how much we don't know. The paradox of learning how to make better decisions is that it requires developing a comfort with doubt.

There are ways, however, of learning to grapple with uncertainty. There are methods for making a vague future more foreseeable by calculating, with some precision, what you do and don't know.

Annie is still alive at the Tournament of Champions. She has enough chips to stay in the game. The dealer gives each player their next cards and another hand begins.

II.

In 2011, the federal Office of the Director of National Intelligence approached a handful of universities with grant money and asked them to participate in a project "to dramatically enhance the accuracy, precision, and timeliness of intelligence forecasts." The idea was that each school would recruit a team of foreign affairs experts, and then ask them to make predictions about the future. Researchers would study who made the most accurate forecasts and, crucially, how they did it. Those insights, the government hoped, would help CIA analysts become better at their jobs.

Most of the universities that participated in the program took a standard approach. They sought out professors, graduate students, international policy researchers, and other specialists. Then they gave them questions no one yet knew the answers to—Will North Korea reenter arms talks by the end of the year? Will the Civic Platform party win the most seats in the Polish parliamentary elections?—and watched how they went about answering. Studying various approaches, everyone figured, would provide the CIA with some fresh ideas.

Two of the universities, however, took a different tack. A group of psychologists, statisticians, and political scientists from the University of Pennsylvania and the University of California–Berkeley, working together, decided to use the government's money as an opportunity to see if they could train regular people to become better forecasters. This group called themselves "the Good Judgment Project," and rather than recruit specialists, the GJP solicited thousands of people—lawyers, housewives, master's students, voracious newspaper readers, and enrolled them in online forecasting classes that taught them different ways of *thinking* about the future. Then, after the training, those participants were asked to answer the same foreign affairs questions as the experts.

For two years, the GJP conducted training sessions, watched people make predictions, and collected data. They tracked who got better, and how performance changed as people were exposed to different types of tutorials. Eventually, the GJP published their findings: Giving participants even brief training sessions in research and statistical techniques—teaching them various ways of thinking about the future—boosted the accuracy of their predictions. And most surprising, a particular kind of lesson—training in how to think probabilistically—significantly increased people's abilities to forecast the future.

The lessons on probabilistic thinking offered by the GJP had instructed participants to think of the future not as what's *going* to happen, but rather as a series of possibilities that *might* occur. It taught them to envision tomorrow as an array of potential outcomes, all of which had different odds of coming true. "Most people are sloppy when they think about the future," said Lyle Ungar, a professor of computer science at the University of Pennsylvania who helped oversee the GJP. "They say things like, 'It's likely we'll go to Hawaii for vacation this year.' Well, does that mean that it's 51 percent certain? Or 90 percent? Because that's a big difference if you're buying nonrefundable tickets." The goal of the GJP's probabilistic training was to show people how to turn their intuitions into statistical estimates.

One exercise, for instance, asked participants to analyze if the French president Nicolas Sarkozy would win reelection in an upcoming vote.

The training indicated that, at a minimum, there were three variables someone should consider in predicting Sarkozy's reelection chances. The first was incumbency. Data from previous French elections indicated that an incumbent such as President Sarkozy, on average, can expect to receive 67 percent of the vote. Based on that, someone might forecast that Sarkozy is 67 percent likely to remain in office.

But there were other variables to take into account, as well. Sarkozy had fallen into disfavor among French voters, and pollsters had estimated that, based on low approval ratings, Sarkozy's reelection chances were actually 25 percent. Under that logic, there was a three-quarters chance he would be voted out. It was also worth considering that the French economy was limping along, and economists guessed that, based on economic performance, Sarkozy would garner only 45 percent of the vote.

So there were three potential futures to consider: Sarkozy could earn 67 percent, 25 percent, or 45 percent of the votes cast. In one scenario, he would win easily, in another he would lose by a wide margin, and the third scenario was a relatively close call. How do you combine those contradictory outcomes into one prediction? "You simply average your estimates based on incumbency, approval ratings, and economic growth rates," the training explained. "If you have no basis for treating one variable as more important than another, use equal weighting. This approach leads you to predict $[(67\% + 25\% + 45\%)/3]$ = approximately a 46% chance of reelection."

3 POSSIBLE FUTURES...

...COMBINED INTO ONE PREDICTION

Nine months later, Sarkozy received 48.4 percent of the vote and was replaced by François Hollande.

This is the most basic kind of probabilistic thinking, a simplistic example that teaches an underlying idea: Contradictory futures can be combined into a single prediction. As this kind of logic gets more sophisticated, experts usually begin speaking about various outcomes as probability curves—graphs that show the distribution of potential futures. For instance, if someone was asked to guess how many seats Sarkozy's party was going to win in the French parliament, an expert might describe the possible outcomes as a curve that shows how the likelihood of winning parliamentary seats is linked to Sarkozy's odds of remaining president:

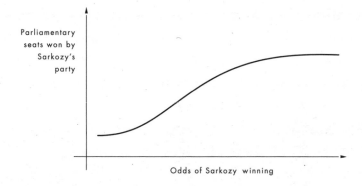

Parliamentary seats won by Sarkozy's party

Odds of Sarkozy winning

In fact, when Sarkozy lost the election, his party, the Union pour un Mouvement Populaire, or UMP, also suffered at the polls, claiming only 194 seats, a significant decline.

The GJP's training modules instructed people in various methods for combining odds and comparing futures. Throughout, a central idea was repeated again and again. The future isn't one thing. Rather, it is a multitude of possibilities that often contradict one another until one of them comes true. And those futures can be combined in order for someone to predict which one is more likely to occur.

This is probabilistic thinking. It is the ability to hold multiple, conflicting outcomes in your mind and estimate their relative likelihoods. "We're not accustomed to thinking about multiple futures,"

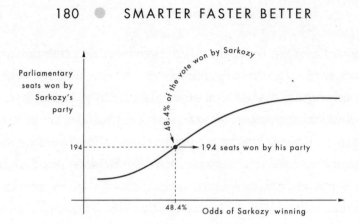

said Barbara Mellers, another GJP leader. "We only live in one reality, and so when we force ourselves to think about the future as numerous possibilities, it can be unsettling for some people because it forces us to think about things we hope won't come true."

Simply exposing participants to probabilistic training was associated with as much as a 50 percent increase in the accuracy of their predictions, the GJP researchers wrote. "Teams with training that engaged in probabilistic thinking performed best," an outside observer noted. "Participants were taught to turn hunches into probabilities. Then they had online discussions with members of their team [about] adjusting the probabilities, as often as every day. . . . Having grand theories about, say, the nature of modern China was not useful. Being able to look at a narrow question from many vantage points and quickly readjust the probabilities was tremendously useful."

Learning to think probabilistically requires us to question our assumptions and live with uncertainty. To become better at predicting the future—at making good decisions—we need to know the difference between what we hope will happen and what is more and less likely to occur.

"It's great to be 100 percent certain you love your girlfriend right now, but if you're thinking of proposing to her, wouldn't you rather know the odds of staying married over the next three decades?" said Don Moore, a professor at UC-Berkeley's Haas School of Business who helped run the GJP. "I can't tell you precisely whether you'll be attracted to each other in thirty years. But I *can* generate some probabilities about the odds of staying attracted to each other, and probabilities about how your goals will coincide, and statistics on how having children might change the relationship, and then you can adjust those likelihoods based on your experiences and what you think is more or less likely to occur, and that's going to help you predict the future a little bit better.

"In the long run, that's pretty valuable, because even though you know with 100 percent certainty that you love her right now, thinking probabilistically about the future can force you to think through things that might be fuzzy today, but are really important over time. It forces you to be honest with yourself, even if part of that honesty is admitting there are things you aren't sure about."

• • •

When Annie started playing poker seriously, it was her brother who sat her down and explained what separated the winners from everyone else. Losers, Howard said, are always looking for certainty at the table. Winners are comfortable admitting to themselves what they *don't* know. In fact, knowing what you don't know is a huge advantage—something that can be used against other players. When Annie would call Howard and complain that she had lost, had suffered bad luck, that the cards had gone against her, he would tell her to stop whining.

"Have you considered that *you* might be the idiot at the table who's looking for certainty?" he asked.

In Texas Hold'Em, the kind of poker Annie was playing, each player received two private cards, and then five communal cards were dealt, faceup, onto the middle of the table to be shared by everyone. The winner was whoever had the best combination of private and communal cards.

When Howard was learning to play, he told Annie, he would go to a late-night game with Wall Street traders, world-champion bridge players, and other assorted math nerds. Tens of thousands of dollars would trade hands as they played until dawn, and then everyone would get breakfast together and deconstruct the games. Howard eventually realized that the hard part of poker wasn't the math. With enough practice, anyone can memorize odds or learn to estimate the chances of winning a pot. No, the hard part was learning to make choices based on probabilities.

For example, let's say you're playing Texas Hold'Em, and you have a queen and nine of hearts as your private cards, and the dealer has put four communal cards on the table:

Communal Cards

One more communal card is going to be dealt. If that last card is a heart, you have a flush, or five hearts, which is a strong hand. A quick mental calculation tells you that since there are 52 cards in a deck, and 4 hearts are already showing, there are 9 possible hearts remaining that might be dealt onto the table, as well as 37 nonheart cards. Put differently, there are 9 cards that will get you a flush, and 37 that won't. The odds, then, of getting a flush are 9 to 37, or roughly 20 percent.*

In other words, there's an 80 percent chance you *won't* make the flush and could lose your money. A novice player, based on those odds, will often fold and get out of the hand. That's because a novice is focused on certainties: The odds of getting a flush are relatively small. Rather than throw money away on an unlikely outcome, they'll quit.

But an expert sees this hand differently. "A good poker player doesn't care about certainty," Annie's brother told her. "They care about knowing what they know and don't know."

For instance, if an expert is holding a queen and nine of hearts and hoping for a flush, and she sees her opponent bet $10, bringing the total pot to $100, a second set of probabilities starts getting calculated. To stay in the game—and see if the last card is a heart—the expert needs only to match the last wager, $10. If the expert bets $10 and makes the flush, she'll win $100. The expert is being offered "pot odds" of 10 to 1, because if she wins, she'll get $10 for every $1 she bets right now.

Now the expert player can compare those odds by imagining this hand one hundred times. The expert doesn't know if she is going to win or lose *this* hand, but she does know that if she played this exact same hand one hundred times, she would, on average, win twenty times, collecting $100 with each victory, yielding $2,000.

---

* Poker is a game of odds within odds. While this example provides an explanation of probabilistic thinking (and the concept of "pot odds"), it is worth noting that a full analysis of this hand is slightly more complex (and would take into account, for instance, the other players at the table). For a more nuanced analysis, please see the notes for chapter 6.

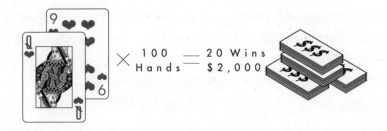

And she knows that playing one hundred times will cost her only an additional $1,000 (because she has to bet only $10 each time). So even if she lost eighty times and won only twenty times, she would still pocket an extra $1,000 (which is the winnings of $2,000 less the $1,000 needed to play).

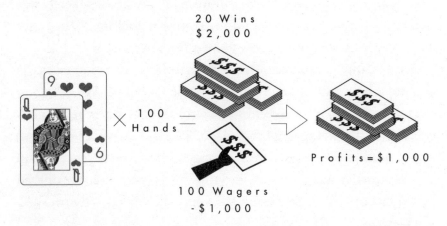

Got it? It's okay if you don't, because the point here is that probabilistic thinking tells the expert how to proceed: She is aware there's a lot she can't predict. But if she played this same hand one hundred times, she would probably end up $1,000 richer. So the expert makes the bet and stays in the game. She knows, from a probabilistic standpoint, it will pay off over time. It doesn't matter that *this* hand is uncertain. What matters is committing to odds that pay off in the long run.

"Most players are obsessed with finding the certainty on the table, and it colors their choices," Annie's brother told her. "Being a great player means embracing uncertainty. As long as you're okay with uncertainty, you can make the odds work for you."

● ● ●

Annie's brother, Howard, is competing in this Tournament of Champions right alongside her when the FossilMan is eliminated. Over the past two decades, Howard has established himself as one of the finest players in the world. He has two World Series of Poker bracelets and millions in winnings. Early in the tournament, Annie and Howard lucked out and didn't have to directly compete for many big pots. Now, however, seven hours have passed.

First the FossilMan was eliminated by that bit of bad luck. Another competitor named Doyle Brunson, a seventy-one-year-old nine-time champion, was knocked out by a risky attempt to double his chips. Phil Ivey, who won his first World Series of Poker tournament at twenty-four, was eliminated by Annie when she drew an ace and queen against Ivey's ace and eight. Over time, the players at the table have dwindled until there are only three players remaining: Annie, Howard, and a man named Phil Hellmuth. It is inevitable Annie and Howard will butt up against each other. The contestants spar over chips and hands for ninety minutes. Then Annie gets a pair of sixes.

She starts tallying what she does and doesn't know. She knows she has strong cards. She knows, from a probabilistic standpoint, that if she played this hand one hundred times, she would do okay. "Sometimes when I'm teaching poker, I'll tell people there are situations where you shouldn't even look at your cards before you bet," Annie told me. "Because if the pot odds are in your favor, you should always make the bet. Just commit to it."

Howard, her brother, seems to like his hand as well, because he pushes all of his chips, $310,000, onto the table. Phil Hellmuth folds. The bet is to Annie.

"I'll call," she says.

They both turn over their cards. Annie reveals her pair of sixes. Howard reveals a pair of sevens.

Annie's Cards   Howard's Cards

"Nice hand, Bub," Annie says. Howard has an 82 percent chance of winning this hand, collecting chips worth more than half a million dollars, and becoming the table's dominant leader. From a probabilistic perspective, they both played this hand exactly right. "Annie made the right choice," Howard later said. "She committed to the odds."

The dealer turns over the first three communal cards.

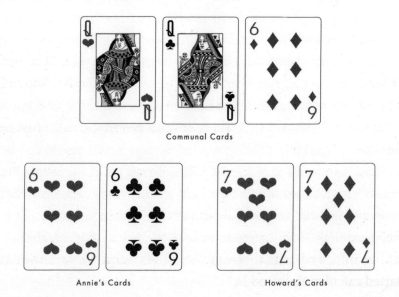

Communal Cards

Annie's Cards   Howard's Cards

"Oh, God," Annie says and covers her face. "Oh, God."

The six and the two queens in the communal pile give Annie a full house. If Annie and Howard replayed this hand one hundred times, Howard would likely win eighty-two of those contests. But not this time. The dealer puts the remaining cards on the table.

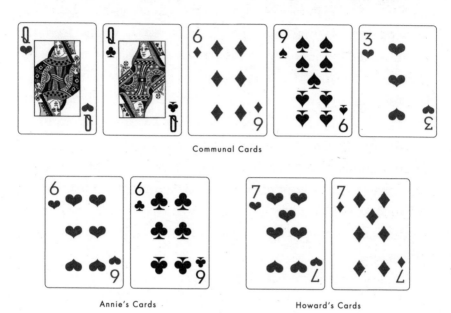

Communal Cards

Annie's Cards   Howard's Cards

Howard is out.

Annie jumps from her chair and hugs her brother. "I'm sorry, Howard," she whispers. Then she runs out of the studio. She starts sobbing before she makes it to the door.

"It's okay," Howard says when he finds her in the hall. "Just beat Phil now."

"You have to learn to live with it," Howard told me later. "I just went through this same thing with my son. He was applying to colleges and he was nervous about it, so we came up with a list of twelve schools—four safety schools, four he had an even chance of getting into, and four that were stretches—and we sat down and started calculating the odds."

By looking at the statistics those schools had published online, Howard and his son calculated the likelihood of getting into each college. Then they added all those probabilities together. It was fairly basic math, the kind even English majors can manage with a little bit of Googling. They figured out that Howard's son had a 99.5 percent chance of getting into at least one school, and a better than even chance of getting into a good school. But it was far from certain he would get into one of the stretch schools, the ones he had fallen in love with. "That was disappointing, but by going through the numbers, he felt less anxious," Howard said. "It prepared him for the possibility that he wouldn't get into his first choice, but he would definitely get in somewhere.

"Probabilities are the closest thing to fortune-telling," Howard said. "But you have to be strong enough to live with what they tell you might occur."

III.

In the late 1990s, a professor of cognitive science at the Massachusetts Institute of Technology named Joshua Tenenbaum began a large-scale examination of the casual ways that people make everyday predictions. There are dozens of questions each of us confront on a daily basis that can be answered only with some amount of forecasting. When we estimate how long a meeting will last, for instance, or envision two driving routes and guess at which one will have less traffic, or predict whether our families will have more fun at the beach or at Disneyland, we're making forecasts that assign likelihoods to various outcomes. We may not realize it, but we're thinking probabilistically. How, Tenenbaum wondered, do our brains do that?

Tenenbaum's specialty was computational cognition—in particular, the similarities in how computers and humans process information. A computer is an inherently deterministic machine. It can predict if your family will prefer the beach or Disneyland only if you give it a specific formula for comparing the merits of beach fun versus amusement parks. Humans, on the other hand, can make such decisions even if we've never visited the seaside or Magic Kingdom before. Our brains can infer from past experiences that, because the kids always complain when we go camping and love watching cartoons, everyone will probably have more fun with Mickey and Goofy.

"How do our minds get so much from so little?" Tenenbaum wrote in a paper published in the journal *Science* in 2011. "Any parent knows, and scientists have confirmed, that typical 2-year-olds can learn how to use a new word such as 'horse' or 'hairbrush' from seeing just a few examples." To a two-year-old, horses and hairbrushes have a great deal in common. The words sound similar. In pictures, they both have long bodies with a series of straight lines—in one case legs, in the other bristles—extruding outward. They come in a range of colors. And yet, though a child might have seen only one picture of a horse and used only one hairbrush, she can quickly learn the difference between those words.

A computer, on the other hand, needs explicit instructions to learn when to use "horse" versus "hairbrush." It needs software that specifies that four legs increases the odds of horsiness, while one hundred bristles increases the probability of a hairbrush. A child can make such calculations before she can form sentences. "Viewed as a computation on sensory input data, this is a remarkable feat," Tenenbaum wrote. "How does a child grasp the boundaries of these subsets from seeing just one or a few examples of each?"

In other words, why are we so good at forecasting certain kinds of things—and thus, making decisions—when we have so little exposure to all the possible odds?

In an attempt to answer this question, Tenenbaum and a colleague, Thomas Griffiths, devised an experiment. They scoured the Internet for data on different kinds of predictable events, such as how much money a movie will make at the box office, or how long the average person lives, or how long a cake needs to bake. They

were interested in these events because if you were to graph multiple examples of each one, a distinct pattern would emerge. Box office totals, for instance, typically conform to a basic rule: There are a
few blockbusters each year that make a huge amount of money, and
lots of other films that never break even.

Within mathematics, this is known as a "power law distribution," and when the revenues of all the movies released in a given
year are graphed together, it looks like this:

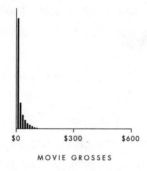

MOVIE GROSSES

Graphing other kinds of events results in different patterns. Take
life spans. A person's odds of dying in a specific year spike slightly
at birth—because some infants perish soon after they arrive—but if
a baby survives its first few years, it is likely to live decades longer.
Then, starting at about age forty, our odds of dying start accelerat-
ing. By fifty, the likelihood of death jumps each year until it peaks at
about eighty-two.

Life spans adhere to a normal, or Gaussian, distribution curve.
That pattern looks like this:

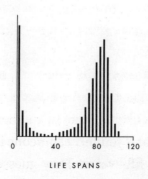

LIFE SPANS

Most people intuitively understand that they need to apply different kinds of reasoning to predicting different kinds of events. We know that box office totals and life spans require different types of estimations, even if we don't know anything about medical statistics or entertainment industry trends. Tenenbaum and Griffiths were curious to find out how people intuitively learn to make such estimations. So they found events with distinct patterns, from box office grosses to life spans, as well as the average length of poems, the careers of congressmen (which adhere to an Erlang distribution), and the length of time a cake needs to bake (which has no strong pattern).

Then they asked hundreds of students to predict the future based on one piece of data:

> You read about a movie that has made $60 million to date. How much will it make in total?
> You meet someone who is thirty-nine years old. How long will he or she live?
> A cake has been baking for fourteen minutes. How much longer does it need to stay in the oven?
> You meet a U.S. congressman who has served for fifteen years. How long will he serve in total?

The students weren't given any additional information. They weren't told anything about power law distributions or Erlang curves. Rather, they were simply asked to make a prediction based on one piece of data and no guidance about what kinds of probabilities to apply.

Despite those handicaps, the students' predictions were startlingly accurate. They knew that a movie that's earned $60 million is a blockbuster, and is likely to take in another $30 million in ticket sales. They intuited that if you meet someone in their thirties, they'll probably live another fifty years. They guessed that if you meet a congressman who has been in power for fifteen years, he'll probably serve another seven or so, because incumbency brings advantages, but even powerful lawmakers can be undone by political trends.

If asked, few of the participants were able to describe the mental logic they used to make their forecasts. They just gave answers that *felt* right. On average, their predictions were often within 10 percent of what the data said was the correct answer. In fact, when Tenenbaum and Griffiths graphed all of the students' predictions for each question, the resulting distribution curves almost perfectly matched the *real* patterns the professors had found in the data they had collected online.

Just as important, each student intuitively understood that different kinds of predictions required different kinds of reasoning. They understood, without necessarily knowing why, that life spans fit into a normal distribution curve whereas box office grosses tend to conform to a power law.

Some researchers call this ability to intuit patterns "Bayesian cognition" or "Bayesian psychology," because for a computer to make those kinds of predictions, it must use a variation of Bayes' rule, a mathematical formula that generally requires running thousands of models simultaneously and comparing millions of results.* At the core of Bayes' rule is a principle: Even if we have very little data, we can still forecast the future by making assumptions and then skewing them based on what we observe about the world. For instance, suppose your brother said he's meeting a friend for dinner. You might forecast there's a 60 percent chance he's going to meet a

---

* Bayes' rule, which was first postulated by the Reverend Bayes in a posthumously published 1763 manuscript, can be so computationally complex that for centuries most statisticians essentially ignored the work because they lacked tools to perform the calculations it demanded. Starting in the 1950s, however, as computers became more powerful, scientists found they could use Bayesian approaches to forecast events that were previously thought unpredictable, such as the likelihood of a war, or the odds that a drug will be broadly effective even if it has only been tested on a handful of people. Even today, though, calculating a Bayesian probability curve can, in some cases, tie up a computer for hours.

man, since most of your brother's friends are male. Now, suppose your brother mentioned his dinner companion was a friend from work. You might want to change your forecast, since you know that most of his coworkers are female. Bayes' rule can calculate the precise odds that your brother's dinner date is female or male based on just one or two pieces of data and your assumptions. As more information comes in—his companion's name is Pat, he or she loves adventure movies and fashion magazines—Bayes' rule will refine the probabilities even more.

Humans can make these kinds of calculations without having to think about them very hard, and we tend to be surprisingly accurate. Most of us have never studied actuarial tables of life spans, but we know, based on experience, that it is relatively uncommon for toddlers to die and more typical for ninety-year-olds to pass away. Most of us don't pay attention to box office statistics. But we are aware that there are a few movies each year that *everyone* sees, and a bunch of films that disappear from the theaters within a week or two. So we make assumptions about life spans and box office revenues based on our experiences, and our instincts become increasingly nuanced the more funerals or movies we attend. Humans are astoundingly good Bayesian predictors, even if we're unaware of it.

Sometimes, however, we make mistakes. For instance, when Tenenbaum and Griffiths asked their students to predict how long an Egyptian pharaoh would reign if he has already ruled for eleven years, a majority of them assumed that pharaohs are similar to other kinds of royalty, such as European kings. Most people know, from reading history books and watching television, that some royalty die early in life. But, in general, if a king or queen survives to middle age, they usually stay on the throne until their hair is gray. So it seemed logical, to Tenenbaum's participants, that pharaohs would be similar. They offered a range of guesses with an average of about twenty-three additional years in power:

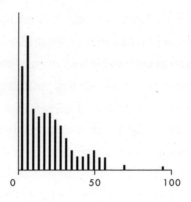

GUESSES ON PHARAOH'S REIGN

That would be a great guess for a British king. But it's a bad guess for an Egyptian pharaoh, because four thousand years ago people had much shorter life spans. Most pharaohs were considered elderly if they made it to thirty-five. So the correct answer is that a pharaoh with eleven years on the throne is expected to reign only another twelve years and then die of disease or some other common cause of death in ancient Egypt:

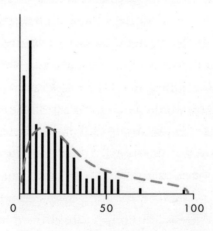

ACTUAL PHARAOH'S REIGN

The students got the reasoning right. They intuited correctly that calculating a pharaoh's reign follows an Erlang distribution. But their assumption—what Bayesians call the "prior" or "base rate"—was off. And because they had a bad assumption about how long ancient Egyptians lived, their subsequent predictions were skewed, as well.

"It's incredible that we're so good at making predictions with such little information and then adjusting them as we absorb data from life," Tenenbaum told me. "But it only works if you start with the right assumptions."

So how do we get the right assumptions? By making sure we are exposed to a full spectrum of experiences. Our assumptions are based on what we've encountered in life, but our experiences often draw on biased samples. In particular, we are much more likely to pay attention to or remember successes and forget about failures. Many of us learn about the business world, for instance, by reading newspapers and magazines. We most frequently go to busy restaurants and see the most popular movies. The problem is that such experiences disproportionately expose us to success. Newspapers and magazines tend to devote more coverage to start-ups that were acquired for $1 billion, and less to the hundreds of similar companies that went bankrupt. We hardly notice the empty restaurants we pass on the way to our favorite, crowded pizza place. We become trained, in other words, to notice success and then, as a result, we predict successful outcomes too often because we're relying on experiences and assumptions that are biased toward all the successes we've seen—rather than the failures we've overlooked.

Many successful people, in contrast, spend an enormous amount of time seeking out information on failures. They read inside the newspaper's business pages for articles on companies that have gone broke. They schedule lunches with colleagues who haven't gotten promoted, and then ask them what went wrong. They request criticisms alongside praise at annual reviews. They scrutinize their credit card statements to figure out why, precisely, they haven't saved as much as they hoped. They pick over their daily missteps when they get home, rather than allowing themselves to forget all the small errors. They ask themselves why a particular call didn't go as well as they had hoped, or if they could have spoken more succinctly at a meeting. We all have a natural proclivity to be optimistic, to ignore our mistakes and forget others' tiny errors. But making good predictions relies on realistic assumptions, and those are based on our experiences. If we pay attention only to good news, we're handicapping ourselves.

"The best entrepreneurs are acutely conscious of the risks that come from only talking to people who have succeeded," said Don Moore, the Berkeley professor who participated in the GJP and who also studies the psychology of entrepreneurship. "They are obsessed with spending time around people who complain about their failures, the kinds of people the rest of us usually try to avoid."

This, ultimately, is one of the most important secrets to learning how to make better decisions. Making good choices relies on forecasting the future. Accurate forecasting requires exposing ourselves to as many successes *and* disappointments as possible. We need to sit in crowded *and* empty theaters to know how movies will perform; we need to spend time around both babies *and* old people to accurately gauge life spans; and we need to talk to thriving *and* failing colleagues to develop good business instincts.

This is hard, because success is easier to stare at. People tend to avoid asking friends who were just fired rude questions; we're hesitant to interrogate divorced colleagues about what precisely went wrong. But calibrating your base rate requires learning from both the accomplished and the humbled.

So the next time a friend misses out on a promotion, ask him why. The next time a deal falls through, call up the other side to find out what you did wrong. The next time you have a bad day or you snap at your spouse, *don't* simply tell yourself that things will go better next time. Instead, force yourself to really figure out what happened.

Then use those insights to forecast more potential futures, to dream up more possibilities of what might occur. You'll never know with 100 percent certainty how things will turn out. But the more you force yourself to envision potential futures, the more you learn about which assumptions are certain or flimsy, the better your odds of making a great decision next time.

● ● ●

Annie knows a lot about Bayesian thinking from graduate school, and she uses it in poker games. "When I play against someone I've never met before, the first thing I do is start thinking about base rates," she told me. "To someone who has never studied Bayes' rule, the way I play might seem like I'm prejudiced, because if I'm sitting across from, say, a forty-year-old businessman, I'm going to assume all he cares about is telling his friends he played against pros and he doesn't really care about winning, so he'll take lots of risks. Or, if I'm sitting across from a twenty-two-year-old in a poker T-shirt, I'm going to assume he learned to play online so he's got a tight, limited game.

"But the difference between prejudice and Bayesian thinking is that I try to *improve* my assumptions as we go along. So once we start playing, if I see that the forty-year-old is a great bluffer, that might mean he's a professional hoping everyone will underestimate him. Or, if the twenty-two-year-old is trying to bluff every hand, it probably means he's some rich kid who doesn't know what he's doing. I spend a lot of time updating my assumptions because, if they're wrong, my base rate is off."

With Annie's brother out of the competition, there are only two players left at the Tournament of Champions table: Annie and Phil Hellmuth. Hellmuth is a card room legend, a television celebrity known as "the Poker Brat." "I'm the Mozart of poker," he told me. "I can read other players probably better than anyone playing, maybe anyone in the world. It's white magic, instinct."

Annie is at one end of the table, Hellmuth at the other. "I had a good idea of how Phil viewed me at that point," Annie said later. "He's told me before that he has a low opinion of my creativity, that he thinks I'm more lucky than smart, that I'm too scared to bluff when it matters."

That's a problem for Annie, because she *wants* Phil to think she's bluffing. The only way she can lure him into a big pot is by convincing him she's bluffing when, in fact, she isn't. To win this tournament, Annie needs to force Phil to change his assumptions of her.

Phil, though, has a different plan. He believes he's the stronger player. He believes he can read Annie. "I have this capacity to learn very, very quickly," he told me. "When I know what people are doing, I can control the table." Those aren't idle boasts. Hellmuth has won fourteen poker championships.

Annie and Phil have roughly equal piles of chips. For the next hour, they play hand after hand, neither gaining a clear advantage. Phil keeps subtly trying to throw Annie off, to make her mad or lose her cool.

"I would have preferred to play your brother," he says.

"This is all right," Annie replies. "I'm just happy to be in the finals."

Annie bluffs Phil four times. "I wanted him to reach the breaking point where he says, 'Screw this, she's bluffing me hand after hand and I gotta fight back,'" Annie said. But Phil doesn't seem shaken. He doesn't overreact.

Finally, Annie gets the hand she's been waiting for. The dealer gives her a king and a nine. Phil receives a king and a seven. In the middle of the table, the dealer lays down a communal king, six, nine, and jack.

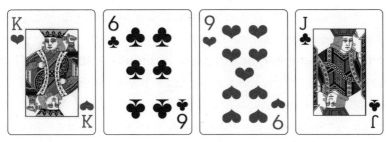

Communal Cards

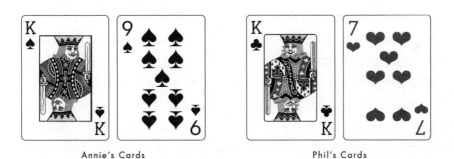

Annie's Cards   Phil's Cards

Phil knows he has a pair of kings. But unbeknownst to him, Annie has *two* pair—kings and nines. Neither sees what the other is holding.

It's Annie's bet, and she raises $120,000. Phil, thinking his pair of kings is likely the strongest hand at the table, matches it. Then Annie goes all in, bringing the pot to $970,000.

The bet is now to Phil.

He starts muttering to himself. "This is unbelievable," he says out loud. "Really unbelievable. She might not even know how strong I am here. I'm not sure she fully even understands the value of the hand."

He stands up.

"I don't know," he says, pacing around the table. "I don't know, I have a bad feeling about this hand." He folds.

Phil flips over his king, showing Annie that he had a pair. Then Annie strikes: She casually turns over one of her cards—but not both—showing Phil her pair of nines, but not revealing that she also had a pair of kings.

"I wanted to force him to change his assumptions about me," Annie later said. "I wanted him to think I was bluffing with a pair of nines."

"Wow, did you really just push in with a nine?" Phil says to Annie. "That's so reckless, especially against someone like me. Maybe I acted too fast."

The players ready for the next hand. Annie has $1,460,000 in chips; Phil has $540,000. The dealer gives them their cards. Annie has a king and a ten; Phil a ten and an eight. The first communal cards come out as a two, ten, and seven.

Communal Cards

Annie's Cards   Phil's Cards

Phil has a pair of tens, with an eight backing it up. It's a good hand. Annie also has a pair of tens, with a king, slightly better.

Phil pushes $45,000 into the pot. Annie raises $200,000. It's an aggressive move. But Phil is starting to believe that Annie is playing recklessly. He thinks he sees a pattern he didn't expect from her: She's bluffing and bluffing and bluffing again. Phil's base rate is gradually shifting.

Phil looks at the pile of chips on the table. Maybe his assumption that Annie is too scared to bluff at critical moments is wrong? Maybe Annie is bluffing right now? Maybe she's finally overplayed her hand?

"I'm all in," Phil says, pushing his stack into the middle of the table.

"I call," Annie says.

Both players turn over their cards.

"Shit," Phil says, seeing that they both have a pair of tens—and that Annie has the high card, a king to Phil's eight.

The dealer puts a seven on the table, benefiting neither player.

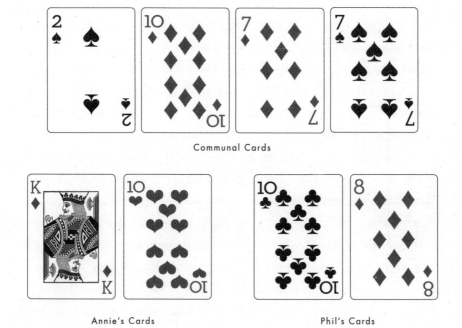

Communal Cards

Annie's Cards   Phil's Cards

Annie is now standing, gripping her cheeks. Phil is also on his feet, breathing hard. "Give me an eight, please," he says. It's the only card that will keep him in the game. The dealer turns over the final communal card. It's a three.

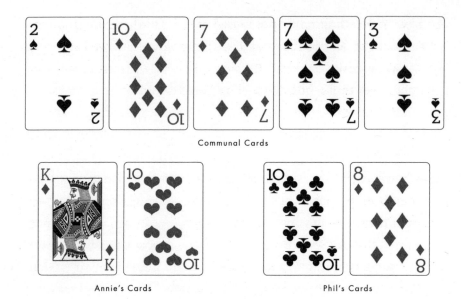

Communal Cards

Annie's Cards   Phil's Cards

Annie wins the $2 million. Phil is out. The game is over. Annie is the champion.

Later, she will tell people that winning this tournament changed her life. It made her, in effect, the most famous female poker player on earth. In 2010, she went on to win the National Heads-Up Poker Championship. Today, she holds a record for World Series of Poker profits. In total, she's won more than $4 million. She doesn't worry about her mortgage anymore. She doesn't have panic attacks. In 2009, she appeared on a season of *Celebrity Apprentice*. She was a little nervous before the filming started, but not too much. There were no anxiety breakdowns. She doesn't play in many poker tournaments these days. She spends most of her time giving lectures to businesspeople about how to think probabilistically, about how to embrace uncertainty, about how, if you commit to a Bayesian outlook, you'll make better decisions in life.

"A lot of poker comes down to luck," Annie told me. "Just like life. You never know where you'll end up. When I checked myself into the psych hospital my sophomore year, there's no way I would have guessed I would end up as a professional poker player. But you have to be comfortable not knowing exactly where life is going. That's how I've learned to keep the anxiety away. All we can do is learn how to make the best decisions that are in front of us, and trust that, over time, the odds will be in our favor."

● ● ●

How do we learn to make better decisions? In part, by training ourselves to think probabilistically. To do that, we must force ourselves to envision various futures—to hold contradictory scenarios in our minds simultaneously—and then expose ourselves to a wide spectrum of successes and failures to develop an intuition about which forecasts are more or less likely to come true.

We can develop this intuition by studying statistics, playing games like poker, thinking through life's potential pitfalls and successes, or helping our kids work through their anxieties by writing them down and patiently calculating the odds. There are numerous ways to build a Bayesian instinct. Some of them are as simple as looking at our past choices and asking ourselves: Why was I so certain things would turn out one way? Why was I wrong?

Regardless of our methods, the goals are the same: to see the future as multiple possibilities rather than one predetermined outcome; to identify what you do and don't know; to ask yourself, which choice gets you the best odds? Fortune-telling isn't real. No one can predict tomorrow with absolute confidence. But the mistake some people make is trying to avoid making any predictions because their thirst for certainty is so strong and their fear of doubt too overwhelming.

If Annie had stayed in academics, would any of this have mattered? "Absolutely," she said. "If you're trying to decide what job to take, or whether you can afford a vacation, or how much you need to save for retirement, those are all predictions." The same basic rules apply. The people who make the best choices are the ones who work hardest to envision various futures, to write them down and think them through, and then ask themselves, which ones do I think are most likely and why?

Anyone can learn to make better decisions. We can all train ourselves to see the small predictions we make every day. No one is right every time. But with practice, we can learn how to influence the probability that our fortune-telling comes true.

7

INNOVATION

How Idea Brokers and Creative Desperation Saved Disney's *Frozen*

The audience starts lining up an hour before the screening room doors open. They are directors and animators, story editors and writers, all of them Disney employees, all eager to see a rough draft of the movie everyone is talking about.

As they settle into their chairs and the lights dim, two sisters appear on the screen against an icy landscape. Anna, the younger character, quickly establishes herself as bossy and uptight, obsessed with her upcoming wedding to the handsome Prince Hans and her coronation as queen. Elsa, her older sister, is jealous, evil—and cursed. Everything she touches turns to ice. She was passed over for the throne because of this power and now, as she runs away from her family to a crystal palace high in the mountains, she nurses a bitter grudge. She wants revenge.

As Anna's wedding day approaches, Elsa plots with a snarky snowman named Olaf to claim the crown for herself. They try to kidnap Anna but their plan is foiled by the square-jawed, dashing

Prince Hans. Bitter Elsa, in a rage, orders an army of snow monsters to descend upon the town and destroy it. The villagers repel the invaders, but when the smoke clears, casualties are discovered: Princess Anna's heart has been partially frozen by her evil sister—and Prince Hans is missing.

The second half of the film follows Anna as she searches for her prince, desperately hoping that his kiss will heal her damaged heart. Meanwhile, Elsa prepares to attack again—and this time floods the village with vicious snow creatures. The monsters, however, are soon out of her control. They begin to threaten everyone, including Elsa herself. The only way to survive, Anna and Elsa realize, is for them to join forces. Through cooperation, they defeat the creatures and the sisters learn that working together is better than struggling apart. They become friends. Anna's heart thaws. Peace returns. Everyone lives happily ever after.

The name of the movie is *Frozen,* and it is scheduled to be released in just eighteen months.

Normally, when a movie screening ends at Disney, there's applause. Often, people cheer or shout. There are usually boxes of tissues inside the screening room because, at Disney, a good cry is the sign of a successful film.

This time, there is no crying. There are no cheers. The tissues go untouched. As everyone files out, they are very, very quiet.

After the screening ended, the film's director, Chris Buck, and about a dozen other Disney filmmakers gathered in one of the studio's dining rooms to discuss what they had just seen. This was a meeting of the studio's "story trust," a group responsible for providing feedback on films as they go through production. As the story trust prepared to discuss the latest draft of *Frozen,* people served themselves from a buffet of Swedish meatballs. Buck didn't get anything to eat. "The last thing I was feeling was hungry," he told me.

Disney's chief creative officer, John Lasseter, kicked off the conversation. "You've got some great scenes here," he said, and mentioned some of the things he particularly liked: The battles were thrilling. The dialogue between the sisters was witty. The snow monsters were terrifying. The film had a good, fast pace. "It's an exciting movie, and the animation is going to be amazing," he said.

And then he began listing the film's flaws. The list was long.

"You haven't dug deep enough," he said after detailing a dozen problems. "There's not enough for the audience to connect with because there's no character to root for. Anna's too uptight and Elsa's too evil. I didn't find myself liking anyone in the movie until the very end."

When Lasseter was done speaking, the rest of the story trust chimed in, pointing out other problems: There were logical holes in the plot—why, for instance, does Anna stick with Prince Hans when he doesn't seem like such a catch? Also, there were too many characters to keep track of. The plot twists were foreshadowed way too much. It didn't seem believable that Elsa would kidnap her sister and then attack the town without trying something less dramatic first. Anna seemed really whiny for someone who lives in a castle, is marrying a prince, and would soon be queen. One member of the story trust—a writer named Jennifer Lee—particularly disliked Elsa's cynical sidekick. "I f'ing hate Olaf," she had scribbled in her notes. "Kill the snowman."

The truth was, Buck wasn't surprised by all the criticisms. His team had sensed the movie wasn't working for months. The film's screenwriter had restructured the script repeatedly, first with Anna and Elsa as strangers rather than sisters, then with Elsa, the cursed sister, assuming the throne and Anna upset at being a "spare, rather than an heir." The songwriters on the film—a husband-and-wife team behind such Broadway hits as *Avenue Q* and *The Book of Mormon*—were exhausted from writing and discarding song after song. They said they couldn't figure out how to make jealousy and revenge into lighthearted themes.

There were versions of the movie where the sisters were normal townspeople rather than royalty, and others where the sisters reconciled over a shared love of reindeer. In one script, they were raised apart. In another, Anna was jilted at the altar. Buck had introduced characters to explain the origins of Elsa's curse, and had tried creating another love interest. Nothing worked. Every time he solved one problem—by making Anna more likable, for instance, or Elsa less bitter—dozens of others popped up.

"Every movie sucks at first," said Bobby Lopez, one of *Frozen*'s songwriters. "But this was like a puzzle where every piece we added upset how everything else fit. And we knew time was running out."

While most animated projects are given four or five years to mature, *Frozen* was on an accelerated schedule. The movie had been in full production for less than a year, but because another Disney movie had recently collapsed, executives had moved *Frozen*'s release date to November 2013, just a year and a half away. "We had to find answers fast," said Peter Del Vecho, the film's producer. "But they couldn't feel clichéd or like a bunch of stories jammed together. The movie had to work emotionally. It was a pretty stressful time."

This conundrum of how to spur innovation on a deadline—or, put another way, how to make the creative process more productive—isn't unique to filmmaking, of course. Every day, students, executives, artists, policy makers, and millions of other people confront problems that require inventive answers delivered as quickly as possible. As the economy changes, and our capacity to achieve creative insights becomes more important than ever, the need for fast originality is even more urgent.

For many people, in fact, figuring out how to accelerate innovation is among their most important jobs. "We're obsessed with the productivity of the creative process," said Ed Catmull, president of Walt Disney Animation Studios and cofounder of Pixar. "We think it's something that can be managed poorly or well, and if we get the creative process right, we find innovations faster. But if we don't manage it right, good ideas are suffocated."

Inside the story trust, the conversation about *Frozen* was winding down. "It seems to me like there's a few different ideas competing inside this movie," Lasseter told Buck, the director. "We've got Elsa's story, we've got Anna's story, and we've got Prince Hans and Olaf the snowman. Each of those stories has great elements. There's a lot of really good material here, but you need to make it into one narrative that connects with the audience. You need to find the movie's core."

Lasseter rose from his seat. "You should take as long as you need to find the answers," he said. "But it would be great if it happened soon."

II.

In 1949, a choreographer named Jerome Robbins contacted his friends Leonard Bernstein and Arthur Laurents with an audacious idea. They should collaborate on a new kind of musical, he told them, modeled on *Romeo and Juliet* but set in modern-day New York City. They could integrate classical ballet with opera and experimental theater, and maybe bring in contemporary jazz and modernist drama, as well. Their goal, Robbins said, should be to establish the avant-garde on Broadway.

Robbins was already famous for creating theater—as well as a life—that pushed boundaries. He was bisexual at a time when homosexuality was illegal. He had changed his name from Jerome Rabinowitz to Jerome Robbins to dodge the anti-Semitism he worried would doom his career. He had named friends as Communists before the House Un-American Activities Committee, terrified that if he didn't cooperate, his sexuality would be publicly revealed and he would be shunned. He was a bully and a perfectionist and so despised by dancers that they sometimes refused to speak to him off the stage. But few refused his invitations to perform. He was widely acknowledged—revered, actually—as one of the most creative artists of his time.

Robbins's *Romeo and Juliet* idea was particularly bold because big Broadway musicals, in those days, tended to adhere to fairly predictable blueprints. Stories were built around a male and a female lead who pushed the plot along with dialogue that was spoken, not sung. There were choruses and dancers, as well as elaborate sets and a few duets about midway through each show. The elements of plot, song, and dance, however, weren't intertwined as they were in, say, ballet, where the story and dancing are one, or opera, where dialogue is sung and music shapes the drama as much as any actor on the stage.

For this new show, Robbins wanted to try something different. "Why couldn't we, in aspiration, try to bring our deepest talents together?" Robbins later said. "Why did Lenny have to write an opera, Arthur a play, me a ballet?" The three men wanted to create something that felt modern yet timeless. When Bernstein and Laurents saw a newspaper article about race riots, they proposed making their musical about two lovers—one Puerto Rican, the other white—whose families were affiliated with warring gangs. The name of the show, they decided, would be *West Side Story*.

Over the next few years the men traded scripts, scores, and choreography ideas. They mailed one another drafts during their long months apart. After half a decade of work, though, Robbins was impatient. This musical was *important,* he wrote to Bernstein and Laurents. It would break new ground. They needed to finish the script. To speed things up, he suggested, they should stop trying to do something new at every turn. Instead, they should stick with conventions they knew, from trial and error, had worked in other shows. But they should combine those conventions in novel ways.

For instance, they had been wrestling with the first meeting between Tony and Maria, the musical's main characters. They should take a page from Shakespeare, Robbins suggested, and have the lovers see each other across a dance floor. But it should be made contemporary, a place where "a wild mambo is in progress with the kids doing all the violent improvisation of jitterbugging."

For the battle in which Tony kills his enemy, Robbins said that the choreography ought to imitate the way battles are staged in motion pictures. "The fight scene must be provoked immediately," Robbins wrote, "or else we're boring the audience." During a dramatic encounter between Tony and Maria, they needed something that echoed the classical marriage scene of *Romeo and Juliet,* but also incorporated the theatricality of opera and a bit of the sentimental romanticism that Broadway audiences loved.

The biggest challenge, however, was figuring out which theatrical conventions were truly powerful and which had become clichés. Laurents, for example, had written a script that was divided into the traditional three acts, but it's "a serious mistake to let the audience out of our grip for two intermissions," Robbins wrote. Motion pictures had proven that you can keep audiences in their seats if the action is always progressing. What's more, Robbins wrote to Laurents, "I like best the sections in which you have gone on your own path, writing in your own style with your own characters and imagination. Least successful are those in which I sense the intimidation of Shakespeare standing behind you." Similarly, roles that were too predictable had to be avoided at all costs. "You are way off the track with the whole character of Anita," Robbins wrote to his colleagues. "She is the typical downbeat blues torch-bearing 2nd character," he remarked. "Forget Anita."

By 1957—eight years after they had first embarked on the project—the men were finally done. They had combined different kinds of theater to create something new: a musical where dance, song, and dialogue were integrated into a story of racism and injustice that was as contemporary as the newspapers sold outside the theater doors. All that was left was to find financial backers. Nearly every producer they approached turned them down. The show was too different from what audiences expected, the moneymen said. Finally, Robbins found financiers willing to support a staging in Washington, D.C.—far enough from Broadway, everyone hoped, that if the show bombed, the news might not spread to New York.

● ● ●

The method Robbins suggested for jump-starting the creative process—taking proven, conventional ideas from other settings and combining them in new ways—is remarkably effective, it turns out. It's a tactic all kinds of people have used to spark creative successes. In 2011, two Northwestern University business school professors began examining how such combinations occur in scientific research. "Combinations of existing material are centerpieces in theories of creativity, whether in the arts, the sciences, or commercial innovation," they wrote in the journal *Science* in 2013. And yet most original ideas grow out of old concepts, and "the building blocks of new ideas are often embodied in existing knowledge." Why are some people so much better at taking those old blocks and stacking them in new ways, the researchers wondered?

The researchers—Brian Uzzi and Ben Jones—decided to focus on an activity they were deeply familiar with: writing and publishing academic papers. They had access to a database of 17.9 million scientific manuscripts published in more than twelve thousand journals. The researchers knew there was no objective way to measure each paper's creativity, but they could *estimate* a paper's originality by analyzing the sources authors had cited in their endnotes. "A paper that combines work by Newton and Einstein is conventional. The combination has happened thousands of times," Uzzi told me. "But a paper that combines Einstein and Wang Chong, the Chinese philosopher, that's much more likely to be creative, because it's such an unusual pairing." Moreover, by focusing primarily on the most popular manuscripts in the database—those studies that had been cited by other researchers thousands of times—they could estimate each manuscript's creative input. "To get into the top 5 percent of the most frequently cited studies, you have to say something pretty new," Uzzi said.

Uzzi and Jones—along with their colleagues Satyam Mukherjee and Mike Stringer—wrote an algorithm to evaluate the 17.9 million papers. By examining how many different ideas each study contained, whether those ideas had been mentioned together previously, and if the papers were popular or ignored, their program could rate each paper's novelty. Then they could look to see if the most creative papers shared any traits.

The analysis told them that some creative papers were short; others were long. Some were written by individuals; the majority were composed by teams. Some studies were authored by researchers at the beginning of their careers; others came from more senior faculty.

In other words, there were lots of different ways to write a creative study.

But almost all of the creative papers had at least one thing in common: They were usually combinations of previously known ideas mixed together in new ways. In fact, on average, 90 percent of what was in the most "creative" manuscripts had already been published elsewhere—and had already been picked over by thousands of other scientists. However, in the creative papers, those conventional concepts were applied to questions in manners no one had considered before. "Our analysis of 17.9 million papers spanning all scientific fields suggests that science follows a nearly universal pattern," Uzzi and Jones wrote. "The highest-impact science is primarily grounded in exceptionally conventional combinations of prior work yet simultaneously features an intrusion of unusual combinations." It was this combination of ideas, rather than the ideas themselves, that typically made a paper so creative and important.

If you consider some of the biggest intellectual innovations of the past half century, you can see this dynamic at work. The field of behavioral economics, which has remade how companies and governments operate, emerged in the mid-1970s and '80s when economists began applying long-held principles from psychology to economics, and asking questions like why perfectly sensible people bought lottery tickets. Or, to cite other juxtapositions of familiar ideas in novel ways, today's Internet social networking companies grew when software programmers borrowed public health models that were originally developed to explain how viruses spread and applying them to how friends share updates. Physicians can now map complicated genetic sequences rapidly because researchers have transported the math of Bayes' rule into laboratories examining how genes evolve.

Fostering creativity by juxtaposing old ideas in original ways isn't new. Historians have noted that most of Thomas Edison's inventions were the result of importing ideas from one area of science into another. Edison and his colleagues "used their knowledge of electromagnetic power from the telegraph industry, where they first worked, to transfer old ideas [to the industries of] lighting, telephone, phonograph, railway and mining," two Stanford professors wrote in 1997. Researchers have consistently found that labs and companies encourage such combinations to spark creativity. A 1997 study of the consumer product design firm IDEO found that most of the company's biggest successes originated as "combinations of existing knowledge from disparate industries." IDEO's designers created a top-selling water bottle, for example, by mixing a standard water carafe with the leak-proof nozzle of a shampoo container.

The power of combining old ideas in new ways also extends to finance, where the prices of stock derivatives are calculated by mixing formulas originally developed to describe the motion of dust particles with gambling techniques. Modern bike helmets exist because a designer wondered if he could take a boat's hull, which can withstand nearly any collision, and design it in the shape of a hat. It even reaches to parenting, where one of the most popular baby books—Benjamin Spock's *The Common Sense Book of Baby and Child Care,* first published in 1946—combined Freudian psychotherapy with traditional child-rearing techniques.

"A lot of the people we think of as exceptionally creative are essentially intellectual middlemen," said Uzzi. "They've learned how to transfer knowledge between different industries or groups. They've seen a lot of different people attack the same problems in different settings, and so they know which kinds of ideas are more likely to work."

Within sociology, these middlemen are often referred to as idea or innovation brokers. In one study published in 2004, a sociologist named Ronald Burt studied 673 managers at a large electronics company and found that ideas that were most consistently ranked as "creative" came from people who were particularly talented at taking concepts from one division of the company and explaining them to employees in other departments. "People connected across groups are more familiar with alternative ways of thinking and behaving," Burt wrote. "The between-group brokers are more likely to express ideas, less likely to have ideas dismissed, and more likely to have ideas evaluated as valuable." They were more credible when they made suggestions, Burt said, because they could say which ideas had already succeeded somewhere else.

"This is not creativity born of genius," Burt wrote. "It is creativity as an import-export business."

What's particularly interesting, however, is that there isn't a specific personality associated with being an innovation broker. Studies indicate that almost anyone can become a broker—as long as they're pushed the right way.

● ● ●

Before rehearsals began for *West Side Story,* Robbins went to his colleagues and said he was dissatisfied with the musical's first scene. As initially envisioned, the show opened in a traditional manner with the play's characters introducing themselves via dialogue that illustrated the plot's central tensions:

ACT 1

SCENE 1

A-rab, a teenager dressed in the uniform of his gang (THE JETS) comes across the stage. Suddenly, two DARK-SKINNED BOYS plummet down from a wall, crashing A-rab to the ground and attacking him. The attackers run off and then several boys—dressed like A-rab—run on from the opposite side.

DIESEL

It's A-rab!

BABY JOHN

He was hit *hard.*

ACTION

An' right on our own turf!

Riff, the leader of THE JETS, enters

RIFF

Straight factualities, A-rab. Who did it?

ACTION

Those buggin' Puerto Ricans!

DIESEL

We're supposed to be the champeens in this area—

MOUTHPIECE

The PR's 're crowdin' *us* like their lousy families 're crowdin' ours!

A-RAB

Let's have some action, Riff.

ACTION

Let's put it on the PRs!

BABY JOHN

A rumble!

RIFF

Whoa, buddy boys! Whadda you diapers know from rumbles? The state of your ignorance is appalling. How do you think the top brass go about a war?

BABY JOHN

Crack-O Jack-O!

RIFF

First—you dispatch scouts to the enemy leader to arrange a war council. Then—

ACTION

Then you go!

RIFF

We oughta get Tony so we can take a vote.

ACTION

He always does what you say anyway. C'mon!

In this version of the opening scene, the audience has learned the basics of the plot within moments of the curtain's rise. They know there are two gangs divided along ethnic lines. They know these gangs are engaged in an ongoing battle. They know there is a hierarchy within each gang—Riff is clearly the leader of the Jets—as well as a certain formality: A rumble can't occur without a meeting of the war council. The audience feels the energy and tension (*Crack-O Jack-O!*) and they learn about another character, Tony, who seems important. All in all, an effective opening.

Robbins discarded it. Too predictable, he said. Lazy and clichéd. Gangs don't just fight, they *own* territory, the same way a dancer *owns* a stage. The opening number of a musical about immigrants and the energy of New York ought to feel ambitious and dangerous—it needed to make the audience feel the same way Robbins, Bernstein, and Laurents had felt when they had come up with this idea. They, the playwrights themselves, were strivers, Robbins told them. They were Jews and outcasts, and this musical was an opportunity to draw on their own experiences of exclusion and ambition and put their own emotions on the stage.

"Robbins could be brutal," said Amanda Vaill, Robbins's biographer. "He could sniff out creative complacency and force people to come up with something newer and better than what everyone else settled for." Robbins was an innovation broker, and he forced everyone around him to become brokers, as well.

This is what appeared onstage—and, later, on movie screens—in what eventually became known as "The *West Side Story* Prologue." It is one of the most influential pieces of theater in the last sixty years:

The opening is musical: half-danced, half-mimed. It is primarily a condensation of the growing rivalry between two teen-age gangs, THE JETS and THE SHARKS, each of which has its own prideful uniform. THE JETS—sideburned, long-haired—are vital, restless, sardonic; The SHARKS are Puerto Ricans.

The stage opens with THE JETS on an asphalt court, snapping their fingers as the orchestra plays. A handball strikes the fence and the music stops. One of the boys, RIFF, indicates with a nod to return the handball to its frightened owner. RIFF's subordinate complies and the music restarts.

THE JETS saunter across the court, and as the music swells, they *pirouette*. They cry "yeah!" and begin a series of *ronds de jambe en l'air*. They own this asphalt. They are poor and ignored by society, but right now, they own this space.

A TEENAGER, the leader of THE SHARKS, appears. THE JETS stop moving. Other SHARKS appear, and they start snapping, and then whirl in a series of *pirouettes* of their own. The SHARKS declare their *own* ownership of the stage.

The two gangs skirmish, contesting territory and dominance, pantomiming threats and apologies, competing but never outright fighting until dozens of SHARKS and JETS are flying across the stage, almost but never touching as they taunt and challenge each other. Then a SHARK trips a JET. The JET pushes his attacker. A cymbal starts chiming and everyone is suddenly atop each other, kicking and punching until a police whistle freezes them and the gangs unite, pretending to be friends in front of OFFICER KRUPKE.

For nine minutes, no dialogue is spoken. Everything is communicated through dance.

The first time *West Side Story* was performed in 1957, the audience wasn't certain what to make of it. The actors were dressed in everyday clothes but moved as if in a classical ballet. The dances were as formalized as *Swan Lake,* but described street fights, an attempted rape, and skirmishes with cops. The music echoed the symphonic tritones of Wagner but also the rhythms of Latin jazz. Throughout the musical, the actors switched between song and dialogue interchangeably.

"The ground rules by which *West Side Story* is played are laid out in the opening number," the theater historian Larry Stempel later wrote. "Before an intelligible sentence has had a chance to be uttered, or a single phrase of music sung, dance has conveyed the essential dramatic information."

When the curtain went down on opening night, there was silence. The audience had just seen a musical about rumbles and murder, songs describing bigotry and prejudice and dances in which hoodlums moved like ballerinas and actors sung slang words with the power of opera stars.

As everyone prepared to take their positions for the curtain call, "we ran to our spaces and faced the audience holding hands. The curtain went up and we looked at the audience, and they looked at us, and we looked at them, and I thought, 'Oh, dear Lord, it's a bomb!'" said Carol Lawrence, who played the original Maria. "And then, as if Jerry had choreographed it, they jumped to their feet. I'd never heard people stamping and yelling, and by that time, Lenny had worked his way backstage, and he came at the final curtain and walked to me, put his arms around me and we wept."

West Side Story went on to become one of the most popular and influential musicals in history. It succeeded by mixing originality and convention to create something new. It took old ideas and put them in novel settings so gracefully that many people never realized they were watching the familiar become unique. Robbins forced his colleagues to become brokers, to put their own experiences onto the stage. "That was a real achievement," Robbins later said.

III.

The space assigned to the *Frozen* team for their daily meetings was large, airy, and comfortable. The walls were covered with sketches of castles and ice caves, friendly-looking reindeer, a snow monster named "Marshmallow," and dozens of concepts for trolls. Each morning at nine A.M., the director, Chris Buck, and his core team of writers and artists would assemble with their coffee cups and to-do lists. The songwriters Bobby Lopez and Kristen Anderson-Lopez would videoconference in from their home in Brooklyn. Then everyone would start panicking about how little time was left.

Anxiety was particularly high the morning after the disastrous screening and meeting with the story trust. From the beginning, the *Frozen* team had known they couldn't simply retell an old fairy tale. They wanted to make a movie that said something new. "It couldn't just be that, at the end, a prince gives someone a kiss and that's the definition of true love," Buck told me. They wanted the film to say something bigger, about how girls don't need to be saved by Prince Charming, about how sisters can save *themselves*. The *Frozen* team wanted to turn the standard princess formula on its head. But that's why they were in such trouble now.

"It was a really big ambition," said Jennifer Lee, who joined the *Frozen* team as a writer after working on another Disney film, *Wreck-It Ralph*. "And it was particularly hard because every movie needs tension, but if the tension in *Frozen* is between the sisters, how do you make them both likable? We tried a jealousy plotline, but it felt petty. We tried a revenge story, but Bobby kept saying we needed an optimistic heroine instead of feuds. The story trust was right: The movie needed to connect *emotionally*. But we didn't know how to get there without falling into clichés."

Everyone in the room was well aware they had only eighteen months to finish the movie. Peter Del Vecho, the producer, asked them all to close their eyes.

"We've tried a lot of different things," he said. "It's okay that we haven't found the answers yet. Every movie goes through this, and every wrong step gets us closer to what works.

"Now, instead of focusing on all the things that aren't working, I want you to think about what could be *right*. I want you to envision your biggest hopes. If we could do anything, what would you want to see on the screen?"

The group sat quietly for a few minutes. Then people opened their eyes and started describing what had excited them about this project in the first place. Some had been drawn to *Frozen* because it offered a chance to upend the way girls are portrayed in films. Others said they were inspired by the idea of a movie where two sisters come together.

"My sister and I fought a lot as kids," Lee told the room. Her parents had divorced when Lee was young. She had eventually moved to Manhattan while her sister became a high school teacher in upstate New York. Then, when Lee was in her early twenties, her boyfriend drowned in a boating accident. Her sister had understood what she was going through at that moment, had been there at a time of need. "There's this moment when you start to see your sibling as a person, instead of a reflection of yourself," Lee told the room. "I think that's what's been bothering me the most about this script. If you have two sisters and one of them is the villain and one is a hero, it doesn't feel real. That doesn't happen in real life. Siblings don't grow apart because one is good and one is bad. They grow apart because they're both messes and then they come together when they realize they need each other. *That's* what I want to show."

Over the next month, the *Frozen* team focused on the relationship between Anna and Elsa, the movie's sisters. In particular, the filmmakers drew on their own experiences to figure out how the siblings related. "We can always find the right story when we start asking ourselves what feels true," Del Vecho told me. "The thing that holds us back is when we forget to use our lives, what's inside our heads, as raw material. That's why the Disney method is so powerful, because it pushes us to dig deeper and deeper until we put ourselves on the screen."

Jerry Robbins pushed his collaborators in *West Side Story* to draw on their own experiences to become creative brokers. The Toyota Production System unlocked employees' capacity to suggest innovations by giving them more control. The Disney system does something different. It forces people to use their own emotions to write dialogue for cartoon characters, to infuse real feelings into situations that, by definition, are unreal and fantastical. This method is worth studying because it suggests a way that anyone can become an idea broker: by drawing on their own lives as creative fodder. We all have a natural instinct to overlook our emotions as creative material. But a key part of learning how to broker insights from one setting to another, to separate the real from the clichéd, is paying more attention to how things make us feel. "Creativity is just connecting things," Apple cofounder Steve Jobs said in 1996. "When you ask creative people how they did something, they feel a little guilty because they didn't really do it, they just saw something. It seemed obvious to them after a while. That's because they were able to connect experiences they've had and synthesize new things. And the reason they were able to do that was that they've had more experiences or they have *thought* more about their experiences than other people." People become creative brokers, in other words, when they learn to pay attention to how things make them react and feel.

"Most people are too narrow in how they think about creativity," Ed Catmull, the president of Disney Animation, told me. "So we spend a huge amount of time pushing people to go deeper, to look further inside themselves, to find something that's real and can be magical when it's put into the mouth of a character on a screen. We all carry the creative process inside us; we just need to be pushed to use it sometimes."

This lesson isn't limited to movies or Broadway. The Post-it note, for instance, was invented by a chemical engineer who, frustrated by bookmarks falling out of his church hymnal, decided to use a new adhesive to make them stay put. Cellophane was developed by an exasperated chemist looking for a way to protect tablecloths from wine spills. Infant formula was created, in part, by an exhausted father who suspended vegetable nutrients in powder so he could feed his crying child in the middle of the night. Those inventors looked to their own lives as the raw materials for innovation. What's notable is that, in each case, they were often in an emotional state. We're more likely to recognize discoveries hidden in our own experiences when necessity pushes us, when panic or frustrations cause us to throw old ideas into new settings. Psychologists call this "creative desperation." Not all creativity relies on panic, of course. But research by the cognitive psychologist Gary Klein indicates that roughly 20 percent of creative breakthroughs are preceded by an anxiety akin to the stress that accompanied *Frozen*'s development, or the pressures Robbins forced onto his *West Side Story* collaborators. Effective brokers aren't cool and collected. They're often worried and afraid.

A few months after the story trust meeting, the songwriters Bobby Lopez and Kristen Anderson-Lopez were walking through Prospect Park in Brooklyn, anxious about all the songs they needed to write, when Kristen asked, "What would it feel like if *you* were Elsa?" As they walked past swingsets and joggers, Kristen and Bobby began discussing what they would do if they were cursed and despised for something they couldn't control. "What if you tried to be good your entire life and it didn't matter because people constantly judged you?" she asked.

Kristen knew this feeling. She had felt other parents' looks when she let their daughters eat ice cream instead of healthy snacks. She'd felt glances when she and Bobby let their girls watch an iPad inside a restaurant because they wanted a moment of peace. Perhaps Kristen wasn't cursed with some deadly power—but she knew what it felt like to be judged. It didn't feel fair. It wasn't her fault that she wanted a career. It wasn't her fault that she wanted to be a good mom *and* be a good wife *and* a successful songwriter, and so, inevitably, that meant things like home-packed snacks and sparkling dinner conversation—not to mention thank-you notes and exercise and replying to emails—sometimes fell by the wayside. She didn't want to apologize for not being perfect. She didn't think she needed to. And she didn't think Elsa should have to apologize for being flawed, either.

"Elsa has tried to do everything right, all her life," Kristen said to Bobby. "Now she's being punished for being herself and the only way out is for her to stop caring, to let it all go."

As they walked, they began riffing, singing snippets of lyrics. What if they wrote a song that started with a fairy-tale opening, Bobby suggested, like the stories they read to their girls at night? Then Elsa could talk about the pressures of being a good girl, said Kristen. She jumped up on a picnic bench. "She could change into a woman," she said. "That's what growing up is, letting go of the things you shouldn't have to care about."

She began singing to an audience of trees and trash cans, trying out lyrics for Elsa to convey that she's done being the good girl, that she doesn't care what anyone thinks anymore. Bobby was recording her impromptu song on his iPhone.

Kristen spread her arms.

Let it go, let it go.
That perfect girl is gone.

"I think you just figured out the chorus," said Bobby.

Back in their apartment, they recorded a rough draft of the song in their makeshift studio. In the background were the clinks of plates from the Greek restaurant downstairs. The next day, they emailed it to Buck, Lee, and the rest of the *Frozen* team. It was part power ballad and part classical aria, but infused with Kristen's and Bobby's frustrations and the emancipation they felt when they let go of people's expectations.

When the *Frozen* team gathered at the Disney headquarters the next morning, they put "Let It Go" on the sound system. Chris Montan, Disney's head of music, slammed his hand on the table.

"That's it," he said. "That's our song. That's what this whole movie is about!"

"I have to go rewrite the beginning of the movie," said Lee.

"I was so happy," Lee told me later. "So relieved. We had struggled for so long, and then we heard 'Let It Go' and, finally, it felt like we had broken through. We could see the movie. We had been carrying the pieces in our heads, but we needed someone to show us ourselves in the characters, to make them familiar. 'Let It Go' made Elsa feel like one of us."

IV.

Seven months later, the *Frozen* team had the first two-thirds of the film figured out. They knew how to make Anna and Elsa likable while driving them apart to create the tension the film needed. They knew how to portray the sisters as hopeful yet troubled. They had even transformed Olaf—the f'ing snowman—into a lovable sidekick. Everything was falling into place.

Except they had no idea how to end the film.

"It was this huge puzzle," said Andrew Millstein, president of Walt Disney Animation Studios. "We tried everything. We knew we wanted Anna to sacrifice herself to save Elsa. We knew we wanted the movie's true love to exist between the sisters. But we had to earn that ending. It had to feel real."

When filmmakers get stuck at Disney, it's referred to as spinning. "Spinning occurs because you're in a rut and can't see your project from different perspectives anymore," said Ed Catmull. So much of the creative process relies on achieving distance, on not becoming overly attached to your creation. But the *Frozen* team had become so comfortable with their vision of the sisters, so relieved to have figured out the movie's basics, so grateful that the creative desperation had lifted a bit, that they had lost their ability to see other paths.

This problem is familiar to anyone who has worked on a long-term creative project. As innovation brokers bring together different perspectives, a creative energy is often released that is heightened by a small amount of tension—such as the pressure that comes from deadlines, or clashes that result when people from different backgrounds meld ideas, or the stresses of collaborators' pushing us to do more. And these "tensions can lead to greater creativity, because all those differences trigger divergent thinking, the ability to see something new when you are forced to look at an idea from someone else's point of view," said Francesca Gino, who studies the psychology of creativity at Harvard Business School. "But when that tension disappears, when you solve the big problem and every-one starts seeing things the same way, people also sometimes start thinking alike and forgetting all the options they have."

The *Frozen* team had solved almost all their problems. No one wanted to lose all the progress they had already made. But they couldn't figure out how to end the film. "You start spinning when your flexibility drops," said Catmull. "You get so devoted to what you've already created. But you have to be willing to kill your dar-lings to go forward. If you can't let go of what you've worked so hard to achieve, it ends up trapping you."

So Disney's executives made a change.

"We had to shake things up," said Catmull. "We had to jolt every-one. So we made Jenn Lee a second director."

In one sense, this change should not have made a huge differ-ence. Lee was already the film's writer. Naming her as a second director, with equal authority to Chris Buck, didn't alter who was participating in the daily conversations. It didn't add any new voices to meetings. And Lee herself was the first to admit that she was as stuck as everyone else.

But, Disney executives hoped, disrupting the team's dynamics just slightly might be enough to stop everyone from spinning in place.

● ● ●

In the 1950s, a biologist named Joseph Connell began traveling between his home in California and the rain forests and coral reefs of Australia in an effort to understand why some parts of the world housed such incredible biological diversity while other regions were so ecologically bland.

Connell had picked Australia for two reasons. First, he hated learning new languages. Second, Australia's forests and seascapes offered perfect examples of biological diversity and homogeneity in close proximity. There were long stretches of the Australian coast where hundreds of different kinds of corals, fish, and sea vegetation lived in very close quarters. Less than a quarter mile away, in another portion of the sea that seemed essentially the same, that diversity would plummet and you might find only one or two kinds of coral and plants. Similarly, some pockets of Australia's rain forests contained dozens of different types of trees, lichen, mushrooms, and vines flourishing side by side. But just a hundred yards away, that would dwindle to just one species of each. Connell wanted to understand why nature's diversity—its capacity for creative origination—was distributed so unevenly.

His quest began in the Queensland rain forests: 12,600 square miles that contain everything from forest canopies to eucalyptus groves, as well as the Daintree tropical forest, where conifers and ferns grow right at the edge of the sea, and the Eungella National Park, where trees are so dense that, at ground level, it can be nearly lightless in the middle of the day. As Connell spent his days walking under green canopies and hacking through thick foliage, he found pockets of biodiversity that seemed to erupt out of nowhere. Then, just a few minutes away, that medley would dwindle until just one or two species remained. What explained this diversity and homogeneity?

Eventually, Connell began noticing something similar at the center of each pocket of biodiversity: There was often evidence that a large tree had fallen. Sometimes he would find a decaying trunk or a deep indentation in the soil. In other verdant pockets, he found charred remains underneath the topsoil, suggesting that a fire—perhaps caused by lightning—had blazed for a brief but intense period before the rain forest's dampness had extinguished the flames.

These fallen trees and fires, Connell came to believe, played a crucial role in allowing species to emerge. Why? Because at some point, there had been a "gap in the forest where the trees had come down or had burned, and that gap was big enough to let sunlight in and allow other species to compete," Connell told me. Retired now, he lives in Santa Barbara, but he remembers the details from those trips. "By the time I found some areas, years had passed since the fire or the tree fall, and so new trees had grown in their place and were blocking out the sun again," he said. "But there had been a time when enough light had made it through that other species were able to claim some territory. There had been some disturbance that had given new plants a chance to compete."

In those regions where trees hadn't fallen or fires hadn't occurred, one species had become dominant and had crowded out all competitors. Put differently, once a species solved the problem of survival, it pushed other alternatives away. But if something altered the ecosystem just a little bit, then biodiversity exploded.

"Only up to a point, though," Connell told me. "If the gap in the forest was *too* big, it had the opposite effect." In those parts of the rain forest where loggers had cleared entire fields, where a huge storm had wiped out whole sections of the forest, or where a fire had spread too far, there was much less diversity, even decades later. If the trauma to the landscape was too great, only the hardiest trees or vines could survive.

Next, Connell looked at reefs along the Australian coast. Here, too, he found a similar pattern. In some places, there was a dizzying assortment of coral and seaweed living in close proximity while, just a few minutes by boat away, one species of fast-growing coral had dominated every square inch. The difference, Connell found, was the frequency and intensity of waves and storms. In those areas with high biodiversity, midsized waves and moderate storms came through occasionally. Alternately, in places with no waves or storms, just a handful of species dominated. Or, when waves were too powerful or storms came through too often, they would scrub the reef clean.

It seemed as if nature's creative capacities depended on some kind of periodic disturbance—like a tree fall or an occasional storm—that temporarily upset the natural environment. But the disturbance couldn't be too small or too big. It had to be just the right size. "Intermediate disturbances are critical," Connell told me.

Within biology, this has become known as the intermediate disturbance hypothesis, which holds that "local species diversity is maximized when ecological disturbance is neither too rare nor too frequent." There are other theories that explain diversification in different ways, but the intermediate disturbance hypothesis has become a staple of biology.

"The idea is that every habitat is colonized by a variety of species, but over time one or a few tend to win out," said Steve Palumbi, the director of Stanford's marine station in Monterey, California. This is called "competitive exclusion." If there are no disturbances to the environment, then the strongest species become so entrenched that nothing else can compete. Similarly, if there are massive, frequent disturbances, only the hardiest species grow back. But if there are intermediate disturbances, then numerous species bloom, and nature's creative capacities flourish.

Human creativity, of course, is different from biological diversity. It's an imprecise analogy to compare a falling tree in the Australian rain forest to a change in management at Disney. Let's play with the comparison for a moment, though, because it offers a valuable lesson: When strong ideas take root, they can sometimes crowd out competitors so thoroughly that alternatives can't prosper. So sometimes the best way to spark creativity is by disturbing things just enough to let some light through.

● ● ●

"The thing I noticed, when I first became a director, was that the change was subtle, but at the same time, very real," Jennifer Lee told me. "When you're a writer, there's certain things you know a film needs, but you're just one voice. You don't want to seem defensive or presumptuous because other people have just as many suggestions and your job is to integrate everyone's ideas.

"A director, though, is in *charge*. So when I became a director, I felt like I had to listen even more closely to what everyone was saying because that was my job now. And as I listened, I started picking up on things I hadn't noticed before."

Some of the animators, for instance, were pushing to use the blizzard at the end of the film as a metaphor for the characters' internal turmoil. Others thought they should withhold any foreshadowing, to make the ending a surprise. As a writer, Lee had viewed those suggestions as devices. But now she understood people were asking for clarity, for a direction in which every choice—from the weather on-screen to choices about what is hidden or revealed— reflected a core idea.

A few months after Lee's promotion, Kristen Anderson-Lopez, the songwriter, sent Lee an email. They had been speaking almost every day for a year at this point. They talked at night and sent each other texts during the day. Their friendship didn't end when Lee became a director. But it changed a little bit.

Kristen was riding a school bus, chaperoning her second-grade daughter on a class trip to the American Museum of Natural History in New York City, when she pulled out her phone and typed a message to Lee.

"Yesterday I went to therapy," she wrote. She and her therapist had discussed the *Frozen* team members' differing opinions about how the movie should end. They had talked about Lee's ascension to director. "I was discussing dynamics and politics and power and all that bullshit and who do you listen to and how do [you] start," she typed. "Then she asked me, 'Why do you do it?'"

"And after parsing out the money and ego stuff, it all really comes down to the fact that I have things I need to share about the human experience," Anderson-Lopez typed. "I want to take what I have learned or felt or experienced and help people by sharing it.

"What is it about this frozen story that you, Bobby and I HAVE to say?" Kristen asked. "For me, it has something to do with not getting frozen in roles that are dictated by circumstances beyond our control."

Lee herself was the perfect example of this. Lee had come to Disney as a new film school graduate with little besides a young daughter and a fresh divorce and student loans, and had quickly become a screenwriter at one of the biggest studios on earth. Now she was the first female director in Disney's history. Kristen and Bobby were examples of people escaping their circumstances, as well. They had fought for years to build the careers they wanted, even when everyone said it was ridiculous to hope they could support themselves by writing songs. Now, here they were, with hit Broadway shows and the life they had always hoped for.

To earn *Frozen*'s ending, Kristen said, they had to find a way to share that sense of possibility with the audience.

"What is it for you?" Kristen typed.

Lee replied twenty-three minutes later. It was seven in the morning in Los Angeles.

"I love your therapist," she wrote, "and you." All the different members of the *Frozen* team had their own ideas for the movie. Everyone on the story trust had become locked into their own concept of how the film should end. But none of them fit together perfectly, Lee felt.

However, *Frozen* could have only one ending. Someone had to make a choice. And the right decision, Lee wrote, is that "fear destroys us, love heals us. Anna's journey should be about learning what love is; it's that simple." At the end of the film, "when she sees her sister out on the fjords, she completes her arc by the ultimate act of true love: sacrificing your needs for someone else's. LOVE is a greater force than FEAR. Go with love."

Becoming a director forced Lee to see things differently—and that small jolt was enough to help her realize what the film needed, and to shift everyone else enough to agree with her.

Later that month, Lee sat down with John Lasseter.

"We need clarity," she told him. "The core of this movie isn't about good and evil, because that doesn't happen in real life. And this movie isn't about love versus hate. That's not why sisters grow apart.

"This is a movie about love and fear. Anna is all about love, and Elsa is all about fear. Anna has been abandoned, so she throws herself into the arms of Prince Charming because she doesn't know the difference between real love and infatuation. She has to learn that love is about sacrifice. And Elsa has to learn that you can't be afraid of who you are, you can't run away from your own powers. You have to embrace your strengths.

"That's what we need to do with the ending, show that love is stronger than fear."

"Say it again," Lasseter told her.

Lee described her theory of love versus fear again, explaining how Olaf, the snowman, embodies innocent love while Prince Hans demonstrates that love without sacrifice isn't really love at all; it's narcissism.

"Say it again," Lasseter said.

Lee said it again.

"Now, go tell the team," said Lasseter.

In June 2013, a few months before the movie was set to open, the *Frozen* team flew to a theater in Arizona to conduct a test screening. What appeared on the screen was completely different from what had been shown in the Disney screening room fifteen months earlier. Anna, the younger sister, was now bubbly, optimistic, and lonely. Elsa was loving but scared of her own powers and tortured by the memory of accidentally injuring her sister when they were young. Elsa runs away to an ice castle, intending to live far from humanity—but she inadvertently plunges her kingdom into an endless winter and partially freezes Anna's heart.

Anna begins searching for a prince in the hope that his true love's kiss will melt the ice in her chest. But the man she finds—Prince Hans—turns out to be intent on taking the throne for himself. Prince Hans imprisons Elsa and abandons the slowly freezing Anna, intent on killing both sisters so he can seize the crown.

Elsa escapes from her cell and, near the end of the movie, is running across the frozen fjords, fleeing the corrupt prince. Anna is growing weaker as the ice inside her chest consumes her heart. A blizzard swirls around the sisters and Hans as they all find one another on the frozen sea. Anna is almost dead from the chill inside her body. Hans raises his sword, ready to slay Elsa and put the throne within his reach. As Hans's blade falls, however, Anna steps in front of the blow. Her body turns to ice just as the sword descends, and it strikes her frozen body rather than her sister. By sacrificing herself, Anna has saved Elsa—and this act of devotion, this genuine demonstration of true love, finally melts Anna's chest. She returns to life, and Elsa, released from the anxiety that she'll hurt the people she loves, can now direct her powers to defeat the evil Hans. She knows now how to end the kingdom's winter. The sisters, united, are powerful enough to overcome their enemies and their self-doubts. Hans is expelled, spring returns, and love defeats fear.

All the elements of a traditional Disney plot were included. There were princesses and ball gowns, a handsome prince, a wisecracking sidekick, and a stream of upbeat songs. But throughout the film, those elements had been disturbed, just enough, to let something new and different emerge. Prince Hans wasn't charming—he was the villain. The princesses weren't helpless; instead, they saved each other. True love didn't arrive in a rescue—rather, it came from siblings learning to embrace their own strengths.

"When did this movie get so good?" Kristen Anderson-Lopez whispered to Peter Del Vecho as the screening ended. *Frozen* would go on to win the Academy Award for Best Animated Feature of 2014. "Let It Go" would win the Academy Award for best original song. The film would become the top-grossing animated movie of all time.

● ● ●

Creativity can't be reduced to a formula. At its core, it needs novelty, surprise, and other elements that cannot be planned in advance to seem fresh and new. There is no checklist that, if followed, delivers innovation on demand.

But the *creative process* is different. We can create the conditions that help creativity to flourish. We know, for example, that innovation becomes more likely when old ideas are mixed in new ways. We know the odds of success go up when brokers—people with fresh, different perspectives, who have seen ideas in a variety of settings—draw on the diversity within their heads. We know that, sometimes, a little disturbance can help jolt us out of the ruts that even the most creative thinkers fall into, as long as those shake-ups are the right size.

If you want to become a broker and increase the productivity of your own creative process, there are three things that can help: First, be sensitive to your own experiences. Pay attention to how things make you think and feel. That's how we distinguish clichés from true insights. As Steve Jobs put it, the best designers are those who "have thought more about their experiences than other people." Similarly, the Disney process asks filmmakers to look inward, to think about their own emotions and experiences until they find answers that make imaginary characters come alive. Jerry Robbins pushed his *West Side Story* collaborators to put their own aspirations and emotions on the stage. Look to your own life as creative fodder, and broker your own experiences into the wider world.

Second, recognize that the panic and stress you feel as you try to create isn't a sign that everything is falling apart. Rather, it's the condition that helps make us flexible enough to seize something new. Creative desperation can be critical; anxiety is what often pushes us to see old ideas in new ways. The path out of that turmoil is to look at what you know, to reinspect conventions you've seen work and try to apply them to fresh problems. The creative pain should be embraced.

Finally, remember that the relief accompanying a creative breakthrough, while sweet, can also blind us to seeing alternatives. It is critical to maintain some distance from what we create. Without self-criticism, without tension, one idea can quickly crowd out competitors. But we can regain that critical distance by forcing ourselves to critique what we've already done, by making ourselves look at it from a completely different perspective, by changing the power dynamics in the room or giving new authority to someone who didn't have it before. Disturbances are essential, and we retain clear eyes by embracing destruction and upheaval, as long as we're sensitive to making the disturbance the right size.

There's an idea that runs through these three lessons: The creative process is, in fact, a process, something that can be broken down and explained. That's important, because it means that anyone can become more creative; we can all become innovation brokers. We all have experiences and tools, disturbances and tensions that can make us into brokers—if, that is, we're willing to embrace that desperation and upheaval and try to see our old ideas in new ways.

"Creativity is just problem solving," Ed Catmull told me. "Once people see it as problem solving, it stops seeming like magic, because it's not. Brokers are just people who pay more attention to what problems look like and how they've been solved before. People who are most creative are the ones who have learned that feeling scared is a good sign. We just have to learn how to trust ourselves enough to let the creativity out."

8

ABSORBING DATA

Turning Information into Knowledge in Cincinnati's Public Schools

Students were settling into their seats as the PA system crackled to life inside South Avondale Elementary School.

"This is Principal Macon," a voice said. "I am declaring a Hot Pencil Drill. Please prepare yourselves, prepare your worksheets, and we will begin in five, four, three, two . . ."

Two minutes and thirty-three seconds later, eight-year-old Dante Williams slammed down his pencil, shot his hand into the air, and twitched impatiently as the teacher scribbled his finish time at the top of the multiplication quiz. Then Dante was out of his chair and flying through the door of his third-grade classroom, arms pumping as he speed-walked down the hallway, his worksheet creased in his fist.

Three years earlier, in 2007, when Dante entered kindergarten, South Avondale had been ranked as one of the worst schools in Cincinnati—which, given that the city had some of the lowest scores in the state, meant that the school was among the worst in

Ohio. That year, South Avondale's students had fared so poorly on their assessment exams that officials declared the school an "academic emergency." Just weeks before Dante had stepped onto campus for the first time, a teenager had been murdered—one bullet to the head, one in the back—right next to South Avondale during a football tournament billed as a "Peace Bowl." That crime, combined with the school's deep dysfunctions, poor academic scores, and a general sense that South Avondale had problems too big for anyone to solve, had caused city officials to ask if the board of education should close the campus altogether. The question, however, was where would they send Dante and his classmates? Nearby schools had scored only slightly better on assessment exams, and if those classrooms were forced to absorb additional kids, they would likely fall apart as well.

The community around South Avondale had been poor for decades. There were race riots in the 1960s, and when the city's factories started closing in the '70s, the area's unemployment had skyrocketed. South Avondale administrators saw students coming to school malnourished and with marks of abuse. In the 1980s, the drug trade around the school exploded and never really let up. At times, the violence got so bad that police would patrol the campus's perimeter while classes were in session. "It could be a pretty scary place," said Yzvetta Macon, who was principal from 2009 to 2013. "Students didn't go to South Avondale unless there was no other place to go."

One thing that wasn't a problem, however, was resources. The city of Cincinnati had poured millions of dollars into South Avondale. Local companies such as Procter & Gamble built computer labs and paid for tutoring and sports programs. In an effort to address the school's shortcomings, city officials spent nearly three times as many dollars on every South Avondale student as they did on students in more affluent communities, such as at the public Montessori campus across town. South Avondale had energetic teachers, devoted librarians and tutors, reading specialists, and guidance counselors who were trained in early childhood education and prepared to help parents sign up for state and federal assistance programs.

The school also used sophisticated software to track students' performance. Administrators had embraced data collection and Cincinnati Public Schools had created an individual website for every South Avondale student—a dashboard of information that detailed kids' attendance, test scores, homework, and classroom participation—that was accessible to parents and educators so they could track who was improving and who was falling behind. The school's faculty received a steady stream of memos and spreadsheets showing how each pupil had fared over the past week, month, and year. South Avondale, in fact, was at the forefront of educational Big Data. "K–12 schools should have a clear strategy for developing a data-driven culture," read a U.S. Department of Education report that helped guide Cincinnati's efforts. By studying each student's statistics closely enough, educators believed they could deliver the specific kind of assistance each kid needed most.

"Any idea or new program, we signed up for," said Elizabeth Holtzapple, director of research and evaluation at Cincinnati Public Schools. "We had seen how data and analytics had turned around other districts, and we were on board."

The turnaround at South Avondale, however, was nowhere to be found. Six years after the online dashboards were introduced, more than 90 percent of South Avondale's teachers admitted they hardly ever looked at them—or used the data sent by the district, or read the memos they received each week. In 2008, 63 percent of South Avondale's third graders failed to meet the state's basic educational benchmarks.

So that year, Cincinnati decided to try something different. The district's top officials targeted South Avondale and fifteen other low-performing campuses in what became known as the "Elementary

Initiative," or EI. The effort was perhaps most notable for what it lacked: The schools were given no additional funds or supplementary teachers; there were no new tutoring sessions or after-school programs; the staff and student body at each campus remained basically the same.

Instead, the EI focused on changing how teachers made decisions in their classrooms. The reforms were built around the idea that data can be transformative, but only if people know how to *use* it. To change students' lives, educators had to understand how to transform all the spreadsheets and statistics and online dashboards into insights and plans. They had to be forced to interact with data until it influenced how they behaved.

By the time Dante entered the third grade, two years after the EI started, the program was already so successful it was hailed by the White House as a model of inner-city reform. South Avondale's test scores went up so much that the school earned an "excellent" rating from state officials. By the end of Dante's third-grade year, 80 percent of his classmates were reading at grade level; 84 percent passed the state math exam. The school had quadrupled the number of students meeting the state's guidelines. "South Avondale drastically improved student academic performance in the 2010–11 academic year and changed the culture of the school," a review by the school district read. The school's transformation was so startling that researchers from around the nation soon began traveling to Cincinnati to figure out what the Elementary Initiative was doing right.

When those researchers visited South Avondale, teachers told them that the most important ingredient in the schools' turnaround was data—the same data, in fact, that the district had been collecting for years. Teachers said that a "data-driven culture" had actually transformed how they made classroom decisions.

When pressed, however, those teachers also said they rarely looked at the online dashboards or memos or spreadsheets the central office sent around. In fact, the EI was succeeding because teachers had been ordered to set aside those slick data tools and fancy software—and were told instead to start manipulating information by hand.

Each school, under orders from the central office, had established a "data room"—in some cases, an empty conference room, in others, a large closet that had previously contained cleaning supplies— where teachers had to transcribe test scores onto index cards. They were told to draw graphs on butcher paper that was taped to walls. They ran impromptu experiments—Do test scores improve if kids are placed in smaller reading groups? What happens when teachers trade off classes?—and then scribbled the results onto whiteboards. Rather than simply receiving information, teachers were forced to *engage* with it. The EI had worked because instead of passively absorbing data, teachers made it "disfluent"—harder to process at first, but stickier once it was really understood. By scribbling out statistics and testing preconceptions, teachers had figured out how to use all the information they were receiving. The Elementary Initiative, paradoxically, had made data more cumbersome to absorb—but more useful. And from those index cards and hand-drawn graphs, better classrooms emerged.

"Something special happened inside those data rooms," said Macon, the principal. South Avondale improved not because teachers had more information but because they learned how to understand it. "With Google and the Internet and all the information we have now, you can find answers to almost anything in seconds," said Macon. "But South Avondale shows there's a difference between finding an answer and understanding what it means."

II.

In the past two decades the amount of information embedded in our daily lives has skyrocketed. There are smartphones that count our steps, websites that track our spending, digital maps to plot our commutes, software that watches our Web browsing, and apps to manage our schedules. We can precisely measure how many calories we eat each day, how much our cholesterol scores have improved each month, how many dollars we spent at restaurants, and how many minutes were allocated to the gym. This information can be incredibly powerful. If harnessed correctly, data can make our days more productive, our diets healthier, our schools more effective, and our lives less stressful.

Unfortunately, however, our ability to learn from information hasn't necessarily kept pace with its proliferation. Though we can track our spending and cholesterol, we still often eat and spend in ways we know we should avoid. Even simple uses of information—such as choosing a restaurant or a new credit card—haven't necessarily become more simple. To find a good Chinese restaurant, is it better to consult Google, ask your Facebook feed, call up a friend, or search your browser history to see where you ordered from last time? To figure out which credit card to sign up for, should you consult an online guide? Call your bank? Open those envelopes piling up on the dining room table?

In theory, the ongoing explosion in information should make the right answers more obvious. In practice, though, being surrounded by data often makes it harder to decide.

This inability to take advantage of data as it becomes more plentiful is called "information blindness." Just as snow blindness refers to people losing the capacity to distinguish trees from hills under a blanket of powder, so information blindness refers to our mind's tendency to stop absorbing data when there's too much to take in.

One study of information blindness was published in 2004 when a group of researchers at Columbia University tried to figure out why some people sign up for 401(k) retirement plans while others don't. They studied almost eight hundred thousand people, across hundreds of companies, who were offered opportunities to enroll in 401(k) plans. For many workers, signing up for the retirement plans should have been an easy choice: The 401(k)s offered large tax savings and many of the companies in the study promised to match employees' contributions—in effect giving them free money. And at firms where workers were offered information on two 401(k) options, 75 percent enrolled. Employees at those companies told researchers that signing up seemed obvious. They looked at the two brochures, picked the plan that seemed most sensible, and then watched their retirement accounts grow fatter over time.

At other companies, even as the number of plans to choose among increased, sign-ups remained high. When workers were offered twenty-five different kinds of plans, 72 percent of them enrolled.

But when employees received information on more than thirty plans, something seemed to change. The amount of information people were receiving became so overwhelming that workers stopped making good choices—or, in some cases, any choice at all. At thirty-nine plans, only 65 percent of people signed up for 401(k) accounts. At sixty plans, participation dropped to 53 percent. "Every ten funds added was associated with 1.5 percent to 2 percent drop in participation," the researchers wrote in their 2004 study. Signing up for a 401(k) was still the right decision. But when information became too plentiful, people put the brochures in a drawer and never looked at them again.

"We've found this in dozens of settings," said Martin Eppler, a professor at the University of St. Gallen in Switzerland who studies information overload. "The quality of people's decisions generally gets better as they receive more relevant information. But then their brain reaches a breaking point when the data becomes too much. They start ignoring options or making bad choices or stop interacting with the information completely."

Information blindness occurs because of the way our brain's capacity for learning has evolved. Humans are exceptionally good at absorbing information—as long as we can break data into a series of smaller and smaller pieces. This process is known as "winnowing" or "scaffolding." Mental scaffolds are like file cabinets filled with folders that help us store and access information when the need arises. If someone is handed a huge wine list at a restaurant, for instance, they'll typically have no problem making a selection because their brain will automatically place what they know about wine into a scaffold of categories they can use to make binary decisions (*Do I want a white or a red? White!*), and then finer subcategories (*Expensive or cheap? Cheap!*) until they confront a final comparison (*The six-dollar Chardonnay or the seven-dollar Sauvignon Blanc?*) that draws upon what they have already learned about themselves (*I like Chardonnay!*). We do this so quickly that, most of the time, we're hardly aware it's occurring.

"Our brains crave reducing things to two or three options," said Eric Johnson, a cognitive psychologist at Columbia University who studies decision making. "So when we're faced with a lot of information, we start automatically arranging it into mental folders and subfolders and sub-subfolders."

This ability to digest large amounts of information by breaking it into smaller pieces is how our brains turn information into knowledge. We learn which facts or lessons to apply in a given situation by learning which folders to consult. Experts are distinguished from novices, in part, by how many folders they carry in their heads. An oenophile will look at a wine list and immediately rely on a vast system of folders—such as vintage and region—that don't occur to novices. The oenophile has learned how to organize information (*Choose the year first, then look at the pricing*) in ways that make it less overwhelming. So while a novice is flipping through pages, the expert is already ignoring whole sections of the wine list.

So when we are presented with information on sixty different 401(k) plans and no obvious way to start analyzing them, our brains pivot to a more binary decision: *Do I try to make sense of all this information, or just stick everything in my drawer and ignore it?*

One way to overcome information blindness is to force ourselves to grapple with the data in front of us, to manipulate information by transforming it into a sequence of questions to be answered or choices to be made. This is sometimes referred to as "creating disfluency" because it relies on doing a little bit of work: Instead of simply choosing the house wine, you have to ask yourself a series of questions (*White or red? Expensive or cheap?*). Instead of sticking all the 401(k) brochures into a drawer, you have to contrast the plans' various benefits and make a choice. It might seem like a small effort at the time, but those tiny bits of labor are critical to avoiding information blindness. The process of creating disfluency can be as minor as forcing ourselves to compare a few pages on a menu, or as big as building a spreadsheet to calculate 401(k) payouts. But regardless of the intensity of the effort, the underlying cognitive activity is the same: We are taking a mass of information and forcing it through a procedure that makes it easier to digest.

"The important step seems to be performing some kind of operation," said Adam Alter, a professor at NYU who has studied disfluency. "If you make people use a new word in a sentence, they'll remember it longer. If you make them write down a sentence with the word, they'll start using it in conversations." When Alter conducts experiments, he sometimes gives people instructions in a hard-to-read font because, as they struggle to make out the words, they read the text more carefully. "The initial difficulty in processing the text leads you to think more deeply about what you're reading, so you spend more time and energy making sense of it," he said. When you ask yourself a few questions about wine, or compare the fees on various 401(k) plans, the data becomes less monolithic and more like a series of decisions. When information is made disfluent, we learn more.

● ● ●

In 1997, executives running the debt collection division of Chase Manhattan Bank began wondering why a particular group of employees in Tampa, Florida, were so much more successful than their peers at convincing people to pay their credit card bills. Chase, at the time, was one of the largest credit card issuers in the nation. As a result, it was also one of the largest debt collectors. It employed thousands of people, in offices all over the country, who sat in cubicles all day and called debtor after debtor, to harass them about overdue credit card bills.

Chase knew from internal surveys that debt collectors didn't especially like their jobs, and executives had grown accustomed to lackluster performance. The company had tried to make the work easier by giving collectors tools to help them convince debtors to pay. As each call occurred, for instance, the computer in front of the debt collector served up information that would assist in tailoring their pitch: It told them the debtor's age, how frequently he or she had paid off their balances, how many other credit cards they owned, what conversational tactics had proven successful in the past. Employees were sent to training sessions and given daily memos with charts and graphs showing the success of various collection tactics.

But almost none of the employees, Chase found, paid much attention to the information they received. No matter how many classes Chase provided or memos they sent, collection rates never seemed to improve much. So executives were pleasantly surprised when one team in Tampa started collecting larger-than-usual amounts.

That group was overseen by a manager named Charlotte Fludd, an evangelical minister in training with a passion for long skirts and Hooters chicken wings, who had started out as a debt collector herself and had worked her way through the ranks until she was overseeing a group responsible for some of the hardest accounts, debtors who were 120 to 150 days overdue. Cardholders that far in arrears almost never paid off their balances. However, Fludd's group was collecting $1 million more per month than any other collection team, even as they were going after some of the most reticent debtors. What's more, Fludd's group reported some of Chase's highest employee satisfaction scores. Even the debtors they collected from, in follow-up surveys, said they had appreciated how they had been treated.

Chase's executives hoped Fludd might share her tactics with other managers, and so they asked her to speak at the company's regional meeting at the Innisbrook Resort near Tampa. The title of her talk was "Optimizing the Mosaix/Voicelink Autodialer System." The room was packed.

"Can you tell us how you schedule your autodialer?" one manager asked.

"Carefully," Fludd said. From 9:15 A.M. to 11:50 A.M., she explained, the collectors called people's home numbers because they were more likely to reach a wife taking care of the kids. Women were more likely to send in a check, Fludd said.

"Then, from noon to one thirty, we call debtors' work numbers,"

Fludd explained, "and we get a lot more men, but you can start the conversation by saying, 'Oh, I'm so glad I caught you on your way to lunch,' like he's real important and his schedule is busy, because that way, he'll want to live up to your expectations and he'll promise to pay.

"Then at dinnertime, we call people we think are unmarried because they're more likely to be lonely and will want to talk, and then right after dinner, we call people whose balances have ballooned up and down, because if they've already had a glass of wine and they're relaxed, we can remind them how good it feels to start paying the card off."

Fludd had dozens of tips like these. She had advice on when to use a comforting tone (if you hear soap operas in the background), when collectors should reveal personal details (if the debtor mentions kids), and when to deploy a stern approach (to anyone invoking religion).

The other managers didn't know what to make of these suggestions. All of them sounded perfectly logical—but they didn't think their employees would be able to use any of them. The average debt collector had just a high school diploma. For many collectors, this was their first full-time job. Managers mostly spent their time reminding employees to avoid sounding so wooden on the phone. Their debt collectors weren't going to be able to pay attention to what television shows were playing in the background or listen for religious references. No one was adept enough at analyzing debtors' records to figure out how to reach a housewife versus her husband. They just talked to whoever picked up the phone. Chase might send the collectors memos each morning, the company might give them computer screens of information and provide them with classes— but managers knew almost no one actually *read* those memos or looked at the screens or used what they learned in class. Simply having a phone conversation with a stranger about a sensitive issue like an overdue bill was overwhelming enough on its own. The average collector couldn't process additional information while conducting a call.

But when Fludd was asked why her employees were so effective at processing more information than the average collector, she didn't have any great answers. She couldn't explain why her workers seemed to absorb so much more. So after the conference, Chase hired the consulting firm Mitchell Madison Group to examine her methods.

"How did you figure out that it's better to call women in the morning?" a consultant named Traci Entel asked her when Fludd was back in the office.

"Do you want me to show you my calendar?" said Fludd. The consultants weren't certain why she needed a calendar to explain her methods, but sure, they said, let's see the calendar. They expected Fludd to pull out a datebook or journal. Instead, she dropped a binder onto the table. Then she wheeled over a cart containing several more binders just like it.

"Okay," Fludd said, leafing through pages filled with numbers and scribbled notes. She found the sheet she was looking for. "One day, I came up with this idea that it would be easier to collect from younger people, because I figured they're more eager to keep a good credit score," she said.

Fludd explained that coming up with such theories was common on her team. Employees would gather during lunch breaks or after work to kick around ideas. Typically, these ideas didn't make much sense—at least, at first. In fact, the ideas were often somewhat nonsensical, such as the suggestion that an irresponsible young person who is already behind on her debts, for some reason, would suddenly care deeply about improving her credit score. But that was okay. The point wasn't to suggest a *good* idea. It was to generate an idea, any idea at all, and then test it.

Fludd looked at her calendar. "So the next day, we started calling people between the ages of twenty-one and thirty-seven." At the end of the shift, employees reported no noticeable change in how much they had convinced people to pay. So the following morning, Fludd changed one variable: She told her employees to call people between the ages of twenty-six and thirty-one. The collection rate improved slightly. The next day, they called a subset of that group, cardholders between twenty-six and thirty-one with balances between $3,000 and $6,000. Collection rates declined. The next day: Cardholders with balances between $5,000 and $8,000. That led to the highest collection rates of the week. In the evenings, before everyone left, managers gathered to review the day's results and speculate on why certain efforts had succeeded or failed. They printed out logs and circled which calls had gone particularly well. That was Fludd's "calendar": the printouts from each day with annotations and employees' comments as well as notes suggesting why certain tactics had worked so well.

With further testing, Fludd determined that her original theory regarding young people was a dud. That, in itself, wasn't surprising. Most of the theories were duds initially. Employees had all kinds of hunches that didn't bear up under testing. But as each experiment unfolded, workers became increasingly sensitive to patterns they hadn't noticed before. They listened more closely. They tracked how debtors would respond to various questions. And eventually, a valuable insight would emerge—like, say, it's better to call people's homes between 9:15 and 11:50 in the morning because the wife will pick up and women are more likely to pay a family's debts. Sometimes, the debt collectors would develop instincts they couldn't exactly put into words but learned to heed nonetheless.

Then someone would propose a new theory or experiment and the process would start all over again. "When you track every call and keep notes and talk about what just happened with the person in the next cubicle, you start paying attention differently," Fludd told me. "You learn to pick up on things."

To the consultants, this was an example of someone using the scientific method to isolate and test variables. "Charlotte's peers would generally change multiple things at once," wrote Niko Cantor, one of the consultants, in a review of his findings. "Charlotte would only change one thing at a time. Therefore she understood the causality better."

But something else was going on, as well. It wasn't just that Fludd was isolating variables. Rather, by coming up with hypotheses and testing them, Fludd's team was heightening their sensitivity to the information flowing past. In a sense, they were adding an element of disfluency to their work, performing operations on the "data" generated during each conversation until lessons were easier to absorb. The spreadsheets and memos that they received each morning, the data that appeared on their screens, the noises they heard in the background of phone calls—that became material for coming up with new theories and running various experiments. Each phone call contained tons of information that most collectors never registered. But Fludd's employees noticed it, because they were looking for clues to prove or disprove theories. They were interacting with the data embodied in each conversation, turning it into something they could use.

This is how learning occurs. Information gets absorbed almost without our noticing because we're so engrossed with it. Fludd took the torrent of data arriving each day and gave her team a method for placing it into folders that made it easier to understand. She helped her employees *do* something with all those memos they received and the conversations they were having—and, as a result, it was easier for them to learn.

III.

Nancy Johnson became a teacher in Cincinnati because she didn't know what else to do with her life. It had taken her seven years to make it through college, and after graduating, she'd become a flight attendant, married a pilot, and then decided to settle down. In 1996, she started substituting in Cincinnati's public schools, hoping it would lead to a full-time job. She went from classroom to classroom, guiding classes on everything from English to biology, until she finally got a permanent offer as a fourth-grade teacher. On her first day, the principal saw her and said, "So *you're* Ms. Johnson." He later admitted he had gotten a number of applications with the same last name and wasn't fully certain which one he had hired.

A few years later, in response to the federal government's No Child Left Behind law, Cincinnati began tracking students' performances in reading and math via standardized exams. Johnson was soon drowning in reports. Each week, she received memos on students' attendance and their progress in vocabulary, math proficiency, reading, writing, literature comprehension, and something called "cognitive manipulation," as well as reviews of her classroom's proficiency, her teaching aptitude, and the school's overall scores. There was so much information that the city had hired a team of data visualization experts to design the weekly memos the district delivered via the Internet dashboards. The graphics team was talented. The charts Johnson received were easy to read, and the Internet sites contained clear summaries and color-coded trend lines.

But in those first few years, Johnson hardly looked at any of it. She was supposed to use all that information in designing her curricula, but it made her head hurt. "There were lots of memos and statistics, and I knew I was supposed to be incorporating them into my classroom, but it all just kind of washed over me," she said. "It felt like there was this gap between all those numbers and what I needed to know to become a better teacher."

Her fourth-grade kids were mostly poor, and many were from single-parent families. She was a good teacher, but her class still fared badly on assessment exams. In 2007, the year before Cincinnati's Elementary Initiative began, her students scored an average of 38 percent proficiency on the state's reading test.

Then, in 2008, the Elementary Initiative was launched. As part of that reform, Johnson's principal mandated that all teachers had to spend at least two afternoons a month in the school's new data room. Around a conference table, teachers were forced to participate in exercises that made data collection and statistical tabulation even *more* time consuming. At the start of the semester, Johnson and her colleagues were told that as part of the EI, they had to create an index card for every student in their class. Then, every other Wednesday, Johnson would go into the data room and transcribe the past two week's test scores onto each student's card, and then group all the cards into color-coded piles—red, yellow, or green—based on whether students were underperforming, meeting expectations, or exceeding their peers. As the semester progressed, she also began grouping cards based on who was improving or falling behind over time.

It was intensely boring. And, frankly, it seemed redundant because all this information was already available on the students' online dashboards. Moreover, many of the people in that room had been teaching for decades. They didn't feel like they needed piles of cards to tell them what was going on in their own classrooms. But an order was an order, and so they went into the data room every other week. "The rule was that everyone had to actually handle the cards, physically move them around," Johnson said. "Everyone hated it, at least at first."

Then one day a third-grade teacher had an idea. Since he had to spend so much time transcribing test scores, he decided to also note on each student's index cards which specific questions they had gotten wrong on that week's assessment exam. He convinced another third-grade teacher to do the same. Next, they combined their cards and made piles by grouping students who had made similar mistakes. When they were done, the piles showed a pattern: A large number of students in one class had done well on pronoun use but had stumbled at fractions; a large number of students in the other classroom had scored the opposite way. The teachers traded curricula. Both classes' scores went up.

The following week, someone else suggested dividing cards from multiple classes into piles based on where students lived. Teachers started giving everyone from the same neighborhoods similar reading assignments. Test scores ticked up. Students were doing their homework together on the bus rides home.

Johnson began putting her students into work groups based on the piles of cards she was making in the data room. Handling the index cards, she found, gave her a more granular sense of each student's strengths and weaknesses. She found herself going into the data room a couple of times a week and putting students' cards into smaller and smaller piles, experimenting with arranging them in different ways. She had felt, before, like she knew her class pretty well. But this was a far deeper level of understanding. "When there are twenty-five students and just one teacher, it's easy to stop seeing them as individuals," she said. "I had always thought of them as a *class*. The data room made me focus on particular kids. It forced me to look at them one by one and ask myself, what does *this* kid need?"

Midway through the year, some of Johnson's colleagues noticed that a small group of students in each class were struggling on math questions. It wasn't a big enough trend that any one teacher would have noticed on their own, but inside the data room, the pattern became clear. That's how the school-wide Hot Pencil Drills started. Soon, students such as eight-year-old Dante were spending each morning filling out multiplication tables as fast as they could, and then speed-walking to the main office to have the fastest test takers' names read over the PA system. Within twelve weeks, the school's math scores were up by 9 percent.

Eight months after the Elementary Initiative was launched, Johnson's class sat for their yearly assessment exam. By that point, she was visiting the data room all the time. She and her colleagues had created dozens of piles of index cards. They had tested various lesson plans and were tracking results on long strips of paper torn from rolls and taped to the walls. Columns of numbers and scribbled notes filled the data room.

The test results came back six weeks later. Johnson's students scored an average of 72 percent, almost double her class's result the previous year. The school's overall scores had more than doubled. In 2009, Johnson became a teacher coach, traveling to other schools in Cincinnati to help instructors learn to use their own data rooms. In 2010, she was selected by her peers as Cincinnati's Educator of the Year.

IV.

Delia Morris was a high school freshman when Cincinnati launched the Elementary Initiative, and so she was too old to benefit from the reforms occurring at places such as South Avondale. And by the time city officials began expanding the program, it seemed too late for her in other respects. Delia's father was fired that year from his job as a security guard at a local grocery store. Then he got into a fight with their landlord. Not long after, Delia came home to find an orange sticker and a padlock on the apartment's front door and everything she and her seven siblings owned stuffed into black garbage bags in the hall. The family was able to stay with people from their church for a while, and then crowded into the apartments of family friends, but from that point on, they moved every few months.

Delia was a good kid and a hard worker. Her teachers had noticed she was unusually smart—gifted enough, they felt, to make it out of Cincinnati's bad neighborhoods and into college. But that didn't mean escape was guaranteed. Every year there were a handful of students who seemed destined for something better until poverty pulled them back down. Delia's teachers were hopeful but not naïve. They knew that even for gifted students, a better life was sometimes out of reach. Delia knew that, as well. She worried that even a whiff of homelessness would change how her teachers perceived her, so she didn't tell anyone what was going on at home. "Going to school was the best part of each day," she told me. "I didn't want to ruin that."

When Delia started her sophomore year at Western Hills High in 2009, the city began expanding its education reforms to high schools. However, some early results among older students proved disappointing. Teachers complained that innovations such as the data rooms were a start but not a solution. Older students were already too hardened, their teachers said; their timelines for intervention were too short. To change kids' lives, they argued, schools needed to help students get better at making the kinds of decisions that offered few opportunities for experimentation. They needed to help teenagers decide between going to college or getting a job; whether to terminate a pregnancy or get married; how to pick among family members when everyone needs your help.

So the school district shifted its focus for high school students. Alongside the Elementary Initiative, the district began creating engineering classes within Western Hills High and other schools in partnership with local universities and the National Science Foundation. The goal was "a multidisciplinary approach to education that encourages students to leverage the technology they use in their daily lives to solve real world problems," a summary of the program read. Ninety percent of students at Western Hills lived below the poverty line. Their classrooms had peeling linoleum floors and cracked chalkboards. "Leveraging technology" was not what most students worried about. Delia signed up for an engineering course taught by Deon Edwards, whose introductory remarks reflected the reality that surrounded all of them.

"We're going to learn how to think like scientists," he told his class. "We're going to leave your parents and friends behind and learn to make choices with clear eyes, without the baggage everyone wants to put on you. And if any of you didn't have anything to eat this morning, I keep energy bars in my desk and you should help yourself. There's nothing wrong with saying you're hungry."

The real focus of Mr. Edwards's class was a system for decision making known as "the engineering design process," which forced students to define their dilemmas, collect data, brainstorm solutions, debate alternative approaches, and conduct iterative experiments. "The engineering design process is a series of steps that engineers follow when they are trying to solve a problem and design a solution for something; it is a methodical approach to problem solving," one teacher's manual explained. The engineering design process was built around the idea that many problems that seem overwhelming at first can be broken into smaller pieces, and then solutions tested, again and again, until an insight emerges. The process asked students to define precisely the dilemma they wanted to solve, then to conduct research and come up with multiple solutions, and then conduct tests, measure results, and repeat the procedure until an answer was found. It told them to make problems more manageable until they fit into scaffolds and mental folders that were easier to carry around.

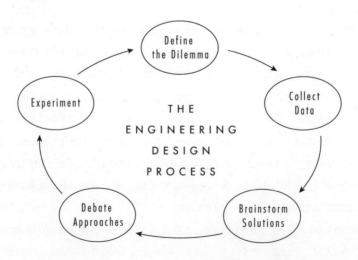

The class's first big assignment was to design an electric car. For weeks, students in Mr. Edwards's class arranged themselves into teams and followed flowcharts detailing each engineering design process step. The classroom had few materials to work with. But that was okay, because the real point of the exercise was to learn how to squeeze information from your environment, no matter where it comes from. Soon students were visiting car dealerships, going to mechanics' shops, and raiding aluminum cans from recycling bins to make battery-testing kits from instructions they had found online. "My first job is to teach them to slow down a little bit," Deon Edwards told me. "These are kids who solve problems all day long. They deal with missing parents and violent boyfriends and classmates on drugs. Everything they experience says they have to choose quickly. I just want to show them that if you have a system for making choices, you can afford to slow down and think."

Midway through the semester, after the class had completed their car designs and moved on to building marble sorters, Delia's twenty-one-year-old sister had a baby. The child's father was out of the picture and Delia's sister, exhausted, begged her to babysit in the afternoons. It felt like a request that was impossible for Delia to refuse. The right decision, Delia's dad told her, was obvious. This was *family*.

So one day in Mr. Edwards's class, Delia pulled the engineering flowchart from her binder and, with her group, put her dilemma through the design process's steps. If she babysat, what would happen? One of the first tasks in engineering design is finding data, so Delia began making a list of experiences that seemed germane. Another sister, Delia told the group, had taken an after-school job a few years earlier and the family had quickly come to rely on that paycheck, making it impossible for her to quit and putting her hopes of community college on hold. If Delia started babysitting, something similar would happen, she suspected. That was data point one.

Then Delia began writing out what her schedule might look like if she was responsible for an infant every afternoon. School from 8:30 to 3:30. Babysitting from 3:30 to 7:30. Homework from 7:30

to 10:00. She would be tired after watching her nephew and would probably end up watching television instead of doing her math or studying for a test. She would become resentful and make bad choices on the weekends. Data point two.

As her group walked through the flowchart, they broke her dilemma into smaller pieces and brainstormed solutions and role-played conversations while the rest of the class discussed how to separate colored marbles from clear ones. Eventually, an answer emerged: Babysitting seemed like a minor sacrifice, but the evidence suggested it wasn't minor at all. Delia prepared a memo for her father listing the steps she had gone through. She wouldn't be able to do it, she told her dad.

Psychologists say learning how to make decisions this way is important, particularly for young people, because it makes it easier for them to learn from their experiences and to see choices from different perspectives. This is a form of disfluency that allows us to evaluate our own lives more objectively, to offset the emotions and biases that might otherwise blind us to the lessons embedded in our pasts. When the animators behind *Frozen* were trying to figure out their film, the Disney system pushed them to look to their own lives as creative fodder. But it's not just creative material we can mine from our experiences—we can find data in our pasts, as well. We all have a natural tendency to ignore the information contained in our previous decisions, to forget that we've already conducted thousands of experiments each time we made a choice. We're often too close to our own experiences to see how to break that data into smaller bits.

But systems such as the engineering design process—which forces us to search for information and brainstorm potential solutions, to look for different kinds of insights and test various ideas—help us achieve disfluency by putting the past in a new frame of reference. It subverts our brain's craving for binary choices—*Should I help my sister or let my family down?*—by learning to reframe decisions in new ways.

One important study of the power of such decision-making frames was published in 1984, after a researcher from Northwestern University asked a group of participants to list reasons why they should buy a VCR based on their own experiences. Volunteers generated dozens of justifications for such a purchase. Some said they felt a VCR would provide entertainment. Others saw it as an investment in their education or a way for their families to spend time together. Then those same volunteers were asked to generate reasons *not* to buy a VCR. They struggled to come up with arguments against the expenditure. The vast majority said they were likely to buy one sometime soon.

Next, the researcher asked a new group of volunteers to come up with a list of reasons *against* purchasing a VCR. No problem, they replied. Some said watching television distracted them from their families. Others said that movies were mindless, and they didn't need the temptation. When those same people were then asked to list reasons *for* buying a VCR, they had trouble coming up with convincing reasons to make the purchase and said they were unlikely to ever buy one.

What interested the researcher was how much each group struggled to adopt an opposing viewpoint once they had an initial frame for making a decision. The two groups were demographically similar. They should have been equally interested in buying a VCR. At the very least, they should have generated equal numbers of reasons to buy or spurn the machines. But once a participant grabbed on to a decision-making frame—*This is an investment in my education* versus *This is a distraction from my family*—they found it hard to envision the choice in a different way. A VCR was either a tool for learning or a time-wasting distraction, based on how the question was framed. Similar results have been found in dozens of other experiments in which people were presented with decisions ranging from the vital, such as end-of-life choices, to the costly, such as buying a car. Once a frame is established, that context is hard to dislodge.

Frames can be uprooted, however, if we force ourselves to seek fresh vantage points. When Delia put her babysitting dilemma through Mr. Edwards's flowcharts, it introduced just enough disfluency to disrupt the frame she had initially assumed she should use. When she went home and walked her father through her logic, it shifted his frame, as well. She couldn't care for her nephew, she told him, because Mr. Edwards's Robotics Club required her to stay at school until six o'clock on Tuesdays and Thursdays, and that club was her path to college. What's more, the other days of the week she needed to get her homework done in the library before coming home because otherwise it wouldn't get finished amid the family's chaos and noise. She reframed the decision as a choice between helping her family now, or succeeding at school and helping in other, more important ways down the road. Her father agreed. They would find another babysitter. Delia needed to stay in school.

"Our brain wants to find a simple frame and stick with it, the same way it wants to make a binary decision," Eric Johnson, the Columbia psychologist, told me. "That's why teenagers get stuck thinking about breaking up with a boyfriend as, 'Do I love him or not?' rather than 'Do I want to be in a relationship, or do I want to be able to leave for college?' Or why, when you're buying a car, you start thinking, 'Do I want the power windows or the GPS?' rather than 'Am I sure I can afford this car?'"

"But when we teach people a process for reframing choices, when we give them a series of steps that causes a decision to seem a little bit different than before," said Johnson, "it helps them take more control of what's going on inside their heads."

One of the best ways to help people cast experiences in a new light is to provide a formal decision-making system—such as a flowchart, a prescribed series of questions, or the engineering design process—that denies our brains the easy options we crave. "Systems teach us how to force ourselves to make questions look unfamiliar," said Johnson. "It's a way to see alternatives."

● ● ●

As Delia moved into her senior year at Western Hills High, her home life became increasingly chaotic. Her sister was there, raising the baby. Another sister had dropped out of school. The family would find a place to live and then something would happen—another lost job or a neighbor who complained about too many people in a one-bedroom apartment—and they would have to move again. In her senior year, Delia's family finally found a long-term rental, but it didn't have heat and, sometimes, when there wasn't money to pay the bill, the electricity went off.

Her teachers, by then, had figured out what was going on, and had seen how hard Delia was working. She was getting straight As. They committed themselves to helping her however they could. When Delia needed to do laundry, her English teacher, Ms. Thole, would invite her over for the afternoon. When Delia seemed exhausted, Mr. Edwards would let her stay late in his classroom and nap, her head on the desk, as he graded exams. They saw her potential. They hoped, with a little help, she could make it to college.

Mr. Edwards, in particular, was a constant in Delia's life. He introduced her to the school's guidance counselor and helped her apply for scholarships. He edited her college applications and made sure they were sent in on time. When Delia had a problem with her friends, when she was fighting with a boyfriend or sparring with her dad, when it seemed like she had too much homework and too little time—whenever it seemed like life was overwhelming—she pulled out Mr. Edwards's flowchart and put her troubles through the engineering design process. It was calming. It helped her think of solutions.

In the spring of Delia's senior year, letters began arriving from scholarship committees. She won the $10,000 Nordstrom Scholarship, then a Rotary prize, then the University of Cincinnati's minority scholar's grant. The envelopes kept coming. Seventeen scholarships in all. She was the class valedictorian and was voted most likely to succeed. The night before graduation, she slept at Ms. Thole's house so she could take a hot shower and curl her hair before the ceremony. In the fall, she enrolled at the University of Cincinnati.

"College is a lot harder than I expected," Delia told me. She's a sophomore now, majoring in information technology. She's often the only girl in her classes and the only black student. The university has tried to help students like Delia, first-generation college attendees, by creating a program named "Gen-1" that provides mentors, tutors, mandatory study sessions, and guidance counseling. Gen-1 participants all live in the same dorm freshman year and sign a seven-page contract in which they promise to abide by a curfew, respect evening quiet hours, and participate in study halls. The idea is to help them get some distance from where they grew up, to see themselves in a new context.

"There's still drama at home," Delia said. But when things feel overwhelming, Delia thinks about Mr. Edwards's class. Any problem can be worked through, step-by-step. "If I take something that's bothering me and make it into smaller pieces, it feels like something I can think about without getting upset," she said.

"I've been through a lot. But I feel like, as long as I've got a system for getting outside my head, I can learn from it. Anything that's happened to me can be a lesson, if I think about it right."

● ● ●

The people who are most successful at learning—those who are able to digest the data surrounding them, who absorb insights embedded in their experiences and take advantage of information flowing past—are the ones who know how to use disfluency to their advantage. They transform what life throws at them, rather than just taking it as it comes. They know the best lessons are those that force us to *do* something and to manipulate information. They take data and transform it into experiments whenever they can. Whether we use the engineering design process or test an idea at work or simply talk through a concept with a friend, by making information more disfluent, we paradoxically make it easier to understand.

In one study published in 2014, researchers from Princeton and UCLA examined the relationship between learning and disfluency by looking at the difference between students who took notes by hand while watching a lecture and those who used laptops. Recording a speaker's comments via longhand is both harder and less efficient than typing on a keyboard. Fingers cramp. Writing is slower than typing, and so you can't record as many words. Students who use laptops, in contrast, spend less time actively working during a lecture, and yet they still collect about twice as many notes as their handwriting peers. Put differently, writing is more disfluent than typing, because it requires more labor and captures fewer verbatim phrases.

When the researchers looked at the test scores of those two groups, however, they found that the hand writers scored twice as well as the typists in remembering what a lecturer said. The scientists, at first, were skeptical. Maybe the hand writers were spending more time studying after class? They conducted a second experiment, but this time they put the laptop users and the hand writers in the same lecture and then took away their notes as soon as it was over, so students couldn't study on their own. A week later, they brought everyone back. Once again, those who took notes by hand scored better on a test of the lecture's content. No matter what constraints were placed on the groups, the students who forced themselves to use a more cumbersome note-taking method—who forced disfluency into how they processed information—learned more.

In our own lives, the same lesson applies: When we encounter new information and want to learn from it, we should force ourselves to *do something* with the data. It's not enough for your bathroom scale to send daily updates to an app on your phone. If you want to lose weight, force yourself to plot those measurements on graph paper and you'll be more likely to choose a salad over a hamburger at lunch. If you read a book filled with new ideas, force yourself to put it down and explain the concepts to someone sitting next to you and you'll be more likely to apply them in your life. When you find a new piece of information, force yourself to engage with it, to use it in an experiment or describe it to a friend—and then you will start building the mental folders that are at the core of learning.

Every choice we make in life is an experiment. Every day offers fresh opportunities to find better decision-making frames. We live in a time when data is more plentiful, cheaper to analyze, and easier to translate into action than ever before. Smartphones, websites, digital databases, and apps put information at our fingertips. But it only becomes useful if we know how to make sense of it.

●●●

In 2013, Dante Williams graduated from the fifth grade at South Avondale Elementary. On his last day of school, he went to a party at the same playground where the teenager had been murdered at the Peace Bowl six years before. There were balloons and a bouncy castle, a cotton candy machine and a DJ. South Avondale was still located inside one of Cincinnati's poorest areas. There were still drugs and boarded-up homes near the campus. But 86 percent of the school's students exceeded the state's education standards that year. The previous year, 91 percent of students had tested above the state's standards. There was a list of kids from outside the district waiting to transfer in.

No school changes because of just one program, of course, just as no student succeeds because of one class or one teacher. Both Dante and Delia, as well as South Avondale and Western Hills High, changed because multiple forces came together at once. There were dedicated teachers and a renewed sense of purpose among administrators. There were focused principals and parents supporting the reforms. But dedication and purpose only succeed when we know how to direct them. The data rooms that turned information into real knowledge, the teachers who learned how to see their students as individuals with different needs and strengths: That's how Cincinnati's public schools shifted.

At the graduation ceremony, as Dante walked across the makeshift stage, his family cheered. Like all diplomas handed out that day, his contained a blank space. There was one last thing, the principal told him. No one was allowed to finish elementary school without doing a final bit of work. Dante had to transform this diploma and make it his own. She handed Dante a pen. He filled in the space with his name.

APPENDIX

A Reader's Guide to Using These Ideas

A few months after I reached out to Atul Gawande—the author and physician from the introduction who helped spark my interest in the science of productivity—I began reporting this book. For almost two years, I conducted interviews with experts, read piles of scientific papers, and tracked down case studies. At some point, I began to imagine that I had become something of a productivity expert myself. When it came time to start writing, I figured, translating all those ideas onto paper would be relatively easy. The words would fly from my fingertips.

That is not what happened.

Some days I would sit at my desk and spend hours jumping from website to website looking for new studies to read, then organize my notes. I would get onto airplanes, my carry-on bag stuffed with scientific papers I intended to read, and spend the flight returning emails, writing to-do lists, and ignoring the big, important tasks I needed to complete.

I had a goal in mind—I wanted to write a book about how we can apply these discoveries in productivity to our own lives—but it seemed so far off, so overwhelming, that I kept focusing on easier-to-accomplish objectives. A few months went by, and all I had to show for it was a series of outlines, but no chapters.

"I feel like a failure," I wrote my editor during one particularly dispiriting moment. "I don't know what I'm doing wrong."

When he wrote back, he pointed out the obvious: Maybe I needed to take what I was learning from the experts and apply it to my own life. I had to live by the principles described in this book.

MOTIVATION

One of my hardest challenges, for instance, concerned my motivation, which seemed to flag at exactly the wrong times. While I was working on this book, I was still also a reporter at *The New York Times*. What's more, I was out promoting my previous book, and trying to be a good father and husband. In other words, I was exhausted. After a long day at the *Times*, I would come home and need to start typing up notes, or draft a chapter, or help put my kids to bed, or clean up the dishes, or reply to emails—and I'd find that self-motivation was in short supply. Emails, in particular, were a small form of daily torture. My in-box was constantly stuffed with questions from colleagues, queries from other authors, correspondence from researchers whom I hoped to interview, and other miscellaneous questions that required a thoughtful response.

However, all I wanted to do was watch TV.

As I struggled each night to find the drive to reply to emails, I began thinking about the key insight from chapter one and the ideas that Gen. Charles Krulak used to redesign Marine Corps boot camp by strengthening recruits' internal locus of control:

- Motivation becomes easier when we transform a chore into a choice. Doing so gives us a sense of control.

On any given day, for instance, I had—let's say—at least about fifty emails that needed responses. Every evening, I would resolve to sit down at my computer and deal with them as soon as dinner was over. And, every evening, I would find ways to procrastinate—by reading the kids one more story, or cleaning up the living room, or checking Facebook—in order to avoid the drudgery of typing response after response. Or, I would sweep through my in-box, hitting the reply key again and again, and then, confronted with a screen full of responses awaiting my words, feel overwhelmed.

General Krulak had told me something that stuck with me: "Most recruits don't know how to force themselves to start something hard. But if we can train them to take the first step by doing something that makes them feel in charge, it's easier to keep going."

I realized that Krulak's insight could help me motivate. And so one night, after putting the kids to bed, I sat at my laptop and hit the reply button, creating a series of responses. Then, as fast as I could, I typed a sentence within each email—any sentence at all—to get me going. For instance, a co-worker sent a note asking if I could join him at a meeting. I had put off replying because I didn't want to attend. I knew the meeting would be long and boring. But I couldn't completely ignore him. So I wrote one sentence in my response:

I can attend, but I'll need to leave after twenty minutes.

I went through two dozen replies just like that, writing a short sentence in each one, hardly thinking about it. And then, I went back and filled in the rest of each email:

Hey Jim,
Sure, I can attend, but I'll need to leave after twenty minutes.
I hope that's okay.

Thanks,
Charles

I noticed two things: First, it was much easier to reply to an email once I had at least one sentence on the screen. Second, and more important, it was easier to get motivated when that first sentence was something that made me feel in control. When I told Jim that I could only stay for twenty minutes, it reminded me that I didn't have to commit to his project if I didn't want to. When I drafted a reply to someone asking me to come speak at a conference, I began by typing:

I would like to leave on Tuesday and be back in New York by Thursday night.

Which reinforced that I was in control of whether I attended or not.

Put differently, as I typed a series of short replies, each reminded me that I was in control of the choices being put before me. (As a psychologist might say, I used those sentences to amplify my internal locus of control.) Within thirty-five minutes, I had cleared out my in-box.

But what about other kinds of procrastination? What about when you're confronting a bigger, more involved task, like writing a long memo or having a hard conversation with a colleague? What if there isn't an easy way to prove to yourself that you're in control? For those, I remember the other key lesson from the motivation chapter:

- Self-motivation becomes easier when we see our choices
  as affirmations of our deeper values and goals.

That's why Marine Corps recruits ask each other "why": "Why are you climbing this mountain?", "Why are you missing the birth of your daughter?", "Why are you cleaning a mess hall, or doing push-ups, or running onto a battlefield when there are safer, easier ways to live?" Forcing ourselves to explain *why* we are doing something helps us remember that this chore is a step along a longer path, and that by choosing to take that journey, we are getting closer to more meaningful objectives.

To motivate myself to read studies on airplanes, for instance, I began writing at the top of each manuscript *why* it was important for me to get that task done. When I pulled a study out of my bag, then, it became a little easier to dive in. Something as simple as jotting down a couple of reasons *why* I am doing something makes it much simpler to start.

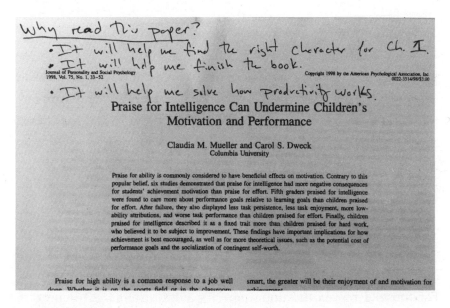

Why read this paper?
• It will help me find the right character for Ch. I.
• It will help me finish the book.
• It will help me solve how productivity works.

Journal of Personality and Social Psychology
1998, Vol. 75, No. 1, 33–52

Praise for Intelligence Can Undermine Children's Motivation and Performance

Claudia M. Mueller and Carol S. Dweck
Columbia University

Praise for ability is commonly considered to have beneficial effects on motivation. Contrary to this popular belief, six studies demonstrated that praise for intelligence had more negative consequences for students' achievement motivation than praise for effort. Fifth graders praised for intelligence were found to care more about performance goals relative to learning goals than children praised for effort. After failure, they also displayed less task persistence, less task enjoyment, more low-ability attributions, and worse task performance than children praised for effort. Finally, children praised for intelligence described it as a fixed trait more than children praised for hard work, who believed it to be subject to improvement. These findings have important implications for how achievement is best encouraged, as well as for more theoretical issues, such as the potential cost of performance goals and the socialization of contingent self-worth.

Praise for high ability is a common response to a job well done. Whether it is on the sports field or in the classroom   smart, the greater will be their enjoyment of and motivation for achievement.

Motivation is triggered by making choices that demonstrate (to ourselves) that we are in control—and that we are moving toward goals that are meaningful. It's that feeling of self-determination that gets us going.

TO GENERATE MOTIVATION

• Make a choice that puts you in control. If you're replying to emails, write an initial sentence that expresses an opinion or decision. If you need to have a hard conversation, decide where it will occur ahead

> of time. The specific choice itself matters less in sparking motivation than the assertion of control.
>
> - Figure out how this task is connected to something you care about. Explain to yourself why this chore will help you get closer to a meaningful goal. Explain *why* this matters—and then, you'll find it easier to start.

GOAL SETTING

Simply figuring out how to motivate myself wasn't always enough, however. Writing a book is a big goal—too big, in many ways, to grasp the entirety of it at first. In trying to figure out how to wrap my head around the objective, I was helped enormously by the reporting I conducted regarding goal setting. The big takeaway was that I needed two kinds of aims:

- I needed a stretch goal, something to spark big ambitions.
- *AND* I needed a SMART goal, to help me form a concrete plan.

One of the most effective ways to formulate both objectives, experts told me, is through a specific kind of to-do list. I needed to write out my goals—but in a way that forced me to identify my stretch objectives and my SMART aims. So I began writing to-do lists, and at the top of each one, I wrote my overarching ambition, what I was working toward in the long term. (That helped me avoid the need for cognitive closure that can force us to become obsessed with short-term, easy-to-achieve goals.) And then underneath, I described a subgoal and all its SMART components, which forced me to come up with a plan—which, in turn, made it more likely that all my goals would be achieved.

One of my stretch goals in reporting this book, for instance, was to find a story that illustrated how mental models worked. I knew that aviation experts felt that mental models played an important role in how pilots responded to emergencies, and so, at the top of my to-do list for this chapter, I wrote:

Ch 3 To-do.

Stretch: Find aviation story (narrowly averted crash?) That demonstrates mental models

Then, below that stretch goal, I wrote my SMART goals related to that big ambition:

Ch 3 To-do

Stretch: Find aviation story (narrowly averted crash?) That demonstrates mental models

Specific: Locate Aviation expert → Papers on Google Scholar

Measurable: Call 4 experts each morning until find right narrative.

Achievable: Clear morning sched; turn off email from 9:00 - 11:30.

Realistic: Monday: 1 hr looking up experts + make call list rank them. By 10:15, start 4 calls; ask for recs

Timeline: 16 calls by Thurs → if no result, new plan. If found story, write and send synopsis to Andy by Fri.

In case that's hard to read, here's what I wrote:

Stretch: Find an aviation story (a narrowly averted crash?) that demonstrates mental models.

Specific: Locate an aviation expert by researching academic papers on Google Scholar.

Measurable: Call four experts each morning until I find the right person/story.

Achievable: Clear my morning schedule to focus on this task, and turn off email from 9:00 to 11:30.

Realistic: On Monday, spend an hour researching aviation experts and creating a call list; rank those experts and, by 10:15, begin my four calls of the day. At the end of each conversation, ask them to recommend other experts to call.

Timeline: If I do four calls a day, then I should have made at least sixteen calls by Thursday. If I haven't found the perfect story by Thursday, I'll come up with a new plan. If I *do* find the right story, I'll send a synopsis to my editor on Friday.

It took only a few minutes to jot down these stretch and SMART goals—but it made a huge difference in how much I got done that week. Now I create a similar to-do list for every big task—and as a result, I know exactly what to do when I sit at my desk each morning. Instead of having to make decisions—and running the risks of distraction—I have a clear sense of how to proceed.

In addition, because I'm always being reminded of my stretch goal, I don't get easily sidetracked, or captured by the need to simply check things off of my list. As scientists might say, I've muted my craving for cognitive closure. I don't stop working merely because I had a good interview, or because I found a helpful study, or because I found an interesting narrative that *might* go in the book. Instead, I'm always reminded that I'm chasing SMART goals for a bigger reason: to find the perfect story, or finish a chapter, or write a book. In fact, I have a whole series of stretch goals to remind me of my grander ambitions:

<u>Ch 3</u>

<u>Stretch:</u> Explain mental models

<u>Stretch:</u> Open w/ AF 447 - explain why it crashed

<u>Stretch:</u> Explain Cog tunnelling

<u>Stretch:</u> Find a plane crash that was averted

<u>Stretch:</u> Find a study that explains mental models in everyday / workplace (?)

TO SET GOALS:

- Choose a stretch goal: an ambition that reflects your biggest aspirations.
- Then, break that into subgoals and develop SMART objectives.

FOCUS

This being real life, however, there are always distractions and other demands competing for my attention. And so, in addition to having a plan, I needed to work on maintaining my focus. There's a key insight from the chapter about the averted aviation disaster of Qantas Flight 32 that I have tried to keep in mind:

• We aid our focus by building mental models—telling our-selves stories—about what we *expect* to see.

To make sure I stayed focused on my stretch and SMART goals, I had to envision what I *expected* to happen when I sat down at my desk each morning. And so, every Sunday night, I got into a habit of taking a few moments with a pad and pen to imagine what the next day and week *ought* to look like. I usually chose three or four things I wanted to make sure happened, and made myself answer a series of questions:

MY GOAL

Find an aviation story that illustrates mental models

WHAT WILL HAPPEN FIRST?

I'll COMPILE A LIST OF AVIATION EXPERTS

↓

WHAT DISTRACTIONS ARE LIKELY TO OCCUR?

THERE WILL BE A TON OF EMAILS WAITING

↓

HOW WILL YOU HANDLE THAT DISTRACTION?

I WON'T CHECK MY EMAIL ACCOUNT UNTIL 11:30

↓

HOW WILL YOU KNOW YOU'VE SUCCEEDED?

I'll HAVE MADE AT LEAST TEN CALLS AND SPOKEN TO FOUR
AVIATION EXPERTS

↓

WHAT IS NECESSARY FOR SUCCESS?

I'LL NEED A CUP OF COFFEE SO I'M NOT TEMPTED TO GET UP

↓

WHAT WILL YOU DO NEXT?

I'LL RESEARCH LEADS AND PREPARE A CALL LIST
FOR THE NEXT DAY

It typically takes only a few minutes to envision what I hope will occur. But by the end of this exercise, I have a story in my mind—a mental model of how my morning should proceed—and, as a result, when distractions inevitably arise, it's easier to decide, in the moment, whether they deserve my focus or can be ignored.

If my email account says there are thirty new messages, I know that I should ignore them until 11:30, because that's what the story inside my head tells me to do. If the phone rings and caller ID indicates it's an expert I'm trying to contact, I'll take the call, because that interruption has a place within my mental model.

I have a stretch goal and a SMART goal that give me a plan—and a picture inside my head of how that plan is supposed to unfold, so making the choices that shape focus is much easier.

TO STAY FOCUSED:

- Envision what will happen. What will occur first? What are potential obstacles? How will you preempt them? Telling yourself a story about what you *expect* to occur makes it easier to decide where your focus should go when your plan encounters real life.

DECISION MAKING

I had worked on coming up with stretch goals and SMART objectives. I had a mental model to stay focused. I had found ways to improve my motivation. Despite all that, however, every so often something came along that would blow my well-crafted intentions apart. Sometimes it was small, like my wife asking if I wanted to get lunch together. Sometimes it was big, like an editor asking me to take on an exciting, but unplanned, assignment.

So how should I make a decision when confronted with the unexpected? Perhaps there was a valuable insight in the chapter on probabilistic thinking:

- Envision multiple futures, and then force myself to figure out which ones are most likely—and why.

For a simple decision like whether I should meet my wife for lunch, the calculus is easy: In one potential future, I take an hour for lunch and come back happy and relaxed. In another, lunch goes long and we spend most of it discussing family logistics and baby-sitter problems, and when I get back to my desk I'm fried—and behind schedule.

By thinking through potential futures, I was better prepared to influence which of those futures would actually occur. When choosing a restaurant to meet my wife, for instance, I suggested one close to my office so I could make it back to my desk quickly. When family logistics came up during lunch, I asked my wife to wait until that evening to talk calendars. By anticipating the future, I was better prepared to make wiser decisions.

But bigger decisions—such as whether to take on an exciting new writing assignment—require a bit more analysis. Midway through writing this book, for instance, a production company asked if I was interested in developing a TV show. To decide if I should pursue that opportunity—which would delay my reporting, but might pay off in the long run—I wrote out a few potential futures of what might happen if I worked on writing a show:

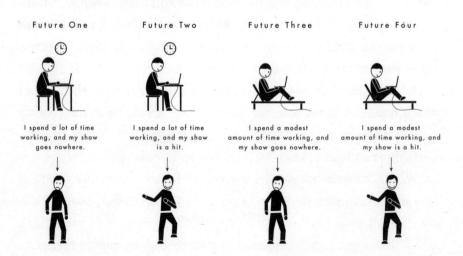

Future One	Future Two	Future Three	Future Four
I spend a lot of time working, and my show goes nowhere.	I spend a lot of time working, and my show is a hit.	I spend a modest amount of time working, and my show goes nowhere.	I spend a modest amount of time working, and my show is a hit.

I had no idea how to evaluate these potential futures. I knew there were dozens of other possibilities I should be considering but couldn't anticipate. And so I called some friends in television. Based on those conversations, I assigned each scenario a rough probability:

Future One	Future Two	Future Three	Future Four
45%	5%	45%	5%
Because you can pour a lot of time into a show, but most never succeed.	Because even though most shows go nowhere, you never know...	Because I can control how much time I invest, if I plan wisely.	Because, who knows?

Based on the estimations of professionals, it seemed most possible that if I invested a lot of time, it wasn't likely to pay off. But if I invested a modest amount of time, there was a likelihood I would learn something, if nothing else.

At that point, I wanted to let my Bayesian instincts guide me, and so I spent a few days letting my imagination play with various outcomes. In the end, I decided that there was another potential future I was ignoring: That even if this show never materialized, I might have a lot of fun. So I decided to commit—but I specified, up front, that I wanted my participation to be modest.

It was a great decision. All told, my involvement in the project was small—probably the equivalent of two weeks. But the payoffs have exceeded my expectations. The show will premiere this fall and I've learned a lot working on it.

What's most important, however, is that I made this decision in a deliberate way. Because I had anticipated various possibilities about what might occur—and, in fact, had drawn up some stretch and SMART goals before joining the project—I was able to manage my involvement.

TO MAKE BETTER DECISIONS:

- Envision multiple futures. By pushing yourself to imagine various possibilities—some of which might be contradictory—you're better equipped to make wise choices.
- We can hone our Bayesian instincts by seeking out different experiences, perspectives, and other people's ideas. By finding information and then letting ourselves sit with it, options become clearer.

THE BIG IDEA

This appendix offers a quick overview of a few key concepts that have been meaningful in my own day-to-day life. If you can become more motivated, more focused, better at setting goals and making good decisions, then you're a long way down the path to becoming more productive. There are, of course, other ideas in this book that also help when we are managing other people, when we are trying to learn faster, when we need to innovate faster. Each of those areas of productivity have their own insights, as well:

TO MAKE TEAMS MORE EFFECTIVE:

- Manage the *how*, not the *who* of teams. Psychological safety emerges when everyone feels like they can speak in roughly equal measure and when teammates show they are sensitive to how each other feel.
- If you are leading a team, think about the message your choices reveal. Are you encouraging equality in speaking, or rewarding the loudest people? Are you showing you are listening by repeating what people say and replying to questions and thoughts? Are you demonstrating sensitivity by reacting when someone seems upset or flustered? Are you showcasing that sensitivity, so other people will follow your lead?

TO MANAGE OTHERS PRODUCTIVELY:

- Lean and agile management techniques tell us employees work smarter and better when they believe they have more decision-making authority *and* when they believe their colleagues are committed to their success.

- By pushing decision making to whoever is closest to a problem, managers take advantage of *everyone*'s expertise and unlock innovation.

- A sense of control can fuel motivation, but for that drive to produce insights and solutions, people need to know their suggestions won't be ignored and that their mistakes won't be held against them.

TO ENCOURAGE INNOVATION:

- Creativity often emerges by combining old ideas in new ways—and "innovation brokers" are key. To become a broker yourself and encourage brokerage within your organization:

- Be sensitive to your own experiences. Paying attention to how things make you think and feel is how we distinguish clichés from real insights. Study your own emotional reactions.

- Recognize that the stress that emerges amid the creative process isn't a sign everything is falling apart. Rather, creative desperation is often critical: Anxiety can be what often pushes us to see old ideas in new ways.

- Finally, remember that the relief accompanying a creative breakthrough, while sweet, can also blind us to alternatives. By forcing ourselves to critique what we've already done, by making ourselves look at it from different perspectives, by giving new authority to someone who didn't have it before, we retain clear eyes.

TO ABSORB DATA BETTER:

- When we encounter new information, we should force ourselves to *do something* with it. Write yourself a note explaining what you just learned, or figure out a small way to test an idea, or graph a series of data points onto a piece of paper, or force yourself to explain an idea to a friend. Every choice we make in life is an experiment—the trick is getting ourselves to see the data embedded in those decisions, and then to use it somehow so we learn from it.

What's most important, throughout all these concepts, is the foundational idea undergirding these lessons, the tissue that connects the eight insights at the heart of this book: Productivity is about recognizing choices that other people often overlook. It's about making certain decisions in certain ways. The way we choose to see our own lives; the stories we tell ourselves, and the goals we push ourselves to spell out in detail; the culture we establish among teammates; the ways we frame our choices and manage the information in our lives. Productive people and companies force themselves to make choices most other people are content to ignore. Productivity emerges when people push themselves to think differently.

When I was working on this book, I came upon a story that I loved, one of my favorite bits of reporting. The tale involved Malcom McLean, the man who essentially created the modern shipping container. McLean died in 2001, but he left behind videotapes and numerous records, and I spent months reading about him, as well as interviewing members of his family and dozens of his former colleagues. They described a man who had relentlessly chased an idea—that shipping goods inside of big metal boxes would make docks more productive—and how that insight eventually transformed manufacturing, the transportation industry, and the economies of whole continents. They explained that McLean was so productive because he was fanatically obsessed with a single idea.

I devoted many, many hours to learning about McLean. I wrote several drafts of his story, determined to fit it into this book.

In the end, however, none of them worked. The lesson he offered—that a single-minded devotion to an idea can spur massive change—turned out not to be as universal and important as the other concepts I wanted to explain. McLean's story was interesting but not vital. What worked for him doesn't work for everyone. There're lots of examples where fanatical devotion has backfired. His insight wasn't big enough to be included among the other eight ideas in this book.

And yet the time I spent researching McLean was worth it, because discarding that work helped me understand the mechanics of focus. My mental model of this book kept conflicting with what I was learning about McLean. My SMART plan for the McLean story didn't match up with my stretch goal of describing universally applicable lessons. In other words, researching McLean helped me figure out what this book was *supposed* to be about. It served as a valuable reminder of how productivity actually functions: Productivity doesn't mean that every action is efficient. It doesn't mean that waste never occurs. In fact, as Disney learned, sometimes you have to foster tension to encourage creativity. Sometimes a misstep is the most important footfall along the path to success.

But in the end, if you learn how to recognize certain choices that, to many, might not be obvious, then you can become smarter, faster, and better over time. Anyone can become more creative, more focused, better at framing their goals and making wise decisions. Schools can be transformed by changing how people absorb data. Teams can be taught how to learn more from mistakes, or use tension to their advantage, or make what seems like misspent hours into lessons getting them closer to their goals. Schools can be remade by empowering the people closest to a problem. The lives of senior citizens can be remade by teaching them to become subversives.

We can all become more productive. Now you know how to start.

ACKNOWLEDGMENTS

The truth of the matter is that most of my own ability to become smarter, faster, and better relies on the kindness of other people, and so there are many of them I wish to thank.

This book exists because Andy Ward willed it into existence by, at first, buying an idea and then, over two years, helping mold it into a book. Everything about Andy—from his graceful editing, to his unyielding demands for quality, to his genuine and heartfelt friendship—inspires the people around him to become better and to want to make the world more beautiful and just. I'm incredibly thankful that I have a chance to know him.

I am also incredibly fortunate to have landed at Random House, which operates under the wise and steady guidance of Gina Centrello, Susan Kamil, and Tom Perry, as well as the superhuman efforts of Maria Braeckel, Sally Marvin, Sanyu Dillon, Theresa Zorro, Avideh Bashirrad, Nicole Morano, Caitlin McCaskey, Melissa Milsten, Leigh Marchant, Alaina Waagner, Dennis Ambrose, Nancy

Delia, Benjamin Dreyer, and the ever-patient Kaela Myers. And I owe a huge debt to all of the people who are so talented at taking these words and putting them into people's hands: David Phethean, Tom Nevins, Beth Koehler, David Weller, Richard Callison, Christine McNamara, Jeffrey Weber, David Romine, Cynthia Lasky, Stacy Berenbaum, Glenn Ellis, Allyson Pearl, Kristen Fleming, Cathy Serpico, Ken Wohlrob, and everyone else in Random House Sales. I am equally lucky to work with Jason Arthur, Emma Finnigan, Matthew Ruddle, Jason Smith, Nigel Wilcockson, and Aslan Byrne at William Heinemann and Martha Konya-Forstner and Cathy Poine in Canada.

I am also indebted to Andrew Wylie and James Pullen at the Wylie Agency. Andrew is steadfast in his desire to make the world safer for his writers, and I am grateful for his efforts. James Pullen has helped me understand how to get published in languages I would have almost certainly failed in high school.

I owe a tremendous amount to *The New York Times*: A huge thanks goes to Dean Baquet, Andy Rosenthal, and Matt Purdy, whose leadership and examples help guide my personal choices daily. Arthur Sulzberger, Mark Thompson, and Meredith Kopit Levien have been great friends and make it possible for the pursuit of truth to occur. I'm so thankful for the time I've spent with Dean Murphy, business editor, and Peter Lattman, deputy business editor, both of whose friendship, advice and patience allowed me to write this book. Similarly, Larry Ingrassia's guidance on nearly every issue is indispensible. Gerry Marzorati has been a great friend, as has Kinsey Wilson, Susan Chira, Jake Silverstein, Bill Wasik, and Cliff Levy.

A few other thanks: I'm indebted to my *Times* colleagues David Leonhardt, A. G. Sulzberger, Walt Bogdanich, Sam Dolnick, Eduardo Porter, David Perpich, Jodi Kantor, Vera Titunik, Peter Lattman, David Segal, Joe Nocera, Michael Barbaro, Jim Stewart, and others who have been so generous with their ideas.

Similarly, I'm thankful to Alex Blumberg, Adam Davidson, Paula

Szuchman, Nivi Nord, Alex Berenson, Nazanin Rafsanjani, Brendan Koerner, Nicholas Thompson, Sarah Ellison, Amanda Schaffer, Dennis Potami, James and Mandy Wynn, Noah Kotch, Greg Nelson, Caitlin Pike, Jonathan Klein, Amanda Klein, Matthew and Chloe Galkin, Nick Panagopulos and Marissa Ronca, Donnan Steele, Stacey Steele, Wesley Morris, Adir Waldman, Rich Frankel, Jennifer Couzin, Aaron Bendikson, Richard Rampell, David Lewicki, Beth Waltemath, Ellen Martin, Amy Wallace, Russ Uman, Erin Brown, Jeff Norton, Raj De Datta, Ruben Sigala, Dan Costello, and Peter Blake, who all provided crucial support and guidance along the way. The book's cover and interior graphics sprung directly from the mind of the incredibly talented Anton Ioukhnovets. Thank you, Anton.

Thank you, as well, to my stalwart fact checkers—Cole Louison and Benjamin Phalen—and Olivia Boone, who helped format and organize the endnotes.

I am indebted to the many people who were generous with their time and knowledge during the reporting of this book. Many are mentioned in the notes, but I wanted to give additional thanks to William Langewiesche, who provided guidance on the mechanics (and writing) of flight, and Ed Catmull and Amy Wallace, who made the Disney chapter happen.

Finally, my deepest thanks are to my family: Katy Duhigg, Jacquie Jenkusky, David Duhigg, Dan Duhigg, Toni Martorelli, Alexandra Alter, and Jake Goldstein have been wonderful friends. My sons, Oliver and Harry, have been sources of inspiration and joy. My parents, John and Doris, encouraged me from a young age to write.

And, of course, my wife, Liz, whose constant love, support, guidance, intelligence, and friendship made this book possible.

—November 2015

A NOTE ON SOURCES

The reporting in this book is based on hundreds of interviews, papers, and studies. Many of those sources are detailed in the text itself or the endnotes, along with guides to additional resources for interested readers.

In most situations, individuals who provided major sources of information or who published research that was integral to reporting were provided with summaries of my reporting and offered the opportunity to review facts and offer additional comments, address discrepancies, or register issues with how information is portrayed. Many of their comments are reproduced in the endnotes. (No source was given access to the book's complete text; all comments are based on summaries provided to sources.) Independent fact-checkers also contacted major sources and reviewed documents to verify and corroborate claims.

In a small number of cases, confidentiality was extended to sources who, for a variety of reasons, did not wish to speak on a for-attribution basis. In three instances, some identifying characteristics have been withheld or slightly modified to conform with patient privacy ethics or for other reasons.

NOTES

CHAPTER ONE: MOTIVATION

11 Ochsner Clinic in New Orleans The facility is now known as the Ochsner Medical Center.

12 *Archives of Neurology* Richard L. Strub, "Frontal Lobe Syndrome in a Patient with Bilateral Globus Pallidus Lesions," *Archives of Neurology* 46, no. 9 (1989): 1024–27.

13 "to get up in the morning" Michel Habib, "Athymhormia and Disorders of Motivation in Basal Ganglia Disease," *The Journal of Neuropsychiatry and Clinical Neurosciences* 16, no. 4 (2004): 509–24.

14 movement and emotion emerge This is how Mauricio Delgado, a neurologist at Rutgers, describes the striatum: "The striatum is the input unit of a larger structure, the basal ganglia. I say the input unit because it receives connections from different brain areas which subserve distinct brain functions—putting the striatum in a prime position to influence behavior. The basal ganglia and in turn the striatum are very important in facets of behavior related to motor (deficits in this structure is common in Parkinson's patients), cognitive and motivation. One line of thinking regarding the striatum and its role in motivation and more specifically reward processing is that it is involved in learning about rewards and using that information to make decisions that help guide behavior, updating the brain along the way whether a reward is better or worse than prior expectations."

14 regulating our moods Oury Monchi et al., "Functional Role of the Basal Ganglia in the Planning and Execution of Actions," *Annals of Neurology* 59, no.2 (2006): 257–64; Edmund T. Rolls, "Neurophysiology and Cognitive Functions of the Striatum," *Revue Neurologique* 150 (1994): 648–60; Patricia S. Goldman-Rakic, "Regional, Cellular, and Subcellular Variations in the Distribution of D_1 and D_5 Dopamine Receptors in Primate Brain," *The Journal of Neuroscience* 15, no. 12 (1995): 7821–36; Bradley Voytek and Robert T. Knight, "Prefrontal Cortex and Basal Ganglia Contributions to Working Memory," *Proceedings of the National Academy of Sciences of the United States of America* 107, no. 42 (2010): 18167–72.

14 motivation had disappeared For my understanding of how brain injuries influence behavior, I am indebted to Julien Bogousslavsky and Jeffrey L. Cummings, *Behavior and Mood Disorders in Focal Brain Lesions* (Cambridge: Cambridge University Press, 2000).

14 striatal injuries Parkinson's frequently involves injuries to the substantia nigra, a region that communicates with the striatum. R. K. B. Pearce et al., "Dopamine Uptake Sites and Dopamine Receptors in Parkinson's Disease and Schizophrenia," *European Neurology* 30, supplement 1 (1990): 9–14; Philip Seeman et al., "Low Density of Dopamine D4 Receptors in Parkinson's, Schizophrenia, and Control Brain Striata," *Synapse* 14, no. 4 (1993): 247–53; Philip Seeman et al., "Human Brain D_1 and D_2 Dopamine Receptors in Schizophrenia, Alzheimer's, Parkinson's, and Huntington's Diseases," *Neuropsychopharmacology* 1, no. 1 (1987): 5–15.

16 see a computer screen Mauricio R. Delgado et al., "Tracking the Hemodynamic Responses to Reward and Punishment in the Striatum," *Journal of Neurophysiology* 84, no. 6 (2000): 3072–77.

16 expectation and excitement In some versions of this experiment, participants were rewarded for guessing right and penalized for guessing wrong with small financial winnings. In response to a fact-checking email, Delgado provided further context for the experiments: "The goal of that initial study was to investigate the human reward circuit. That is, we know from animal research that certain brain regions were important for processing information about reward. We knew less about how that translated to the human brain and how it translated to more common human rewards such as money, which had implications to behavioral addictions such as pathological gambling. Thus, with the guessing game, our initial goal was to compare what happened in the brain when participants received a monetary reward (for a correct guess) and a monetary punishment or loss (for an incorrect guess). The pattern we observe is very characteristic of a reward response. We see activity in the striatum (both dorsal and ventral parts). The response is an initial increase at the beginning of the trial when the question mark appears and they make a guess. We reasoned that it reflected anticipation of a potential reward. Other work using this task (see Delgado et al. 2004, Leotti and Delgado 2011) support that as does the work by Brian Knutson (2001). They don't know yet if their guess is correct and lead to a reward or incorrect and lead to a loss. So the increase is common for both types of trials. Once the outcome is revealed, we see an interesting pattern where the striatum differentiates between a positive and negative outcome—a gain or a loss. It is increased for a gain and decreased response for a loss. One interpretation of this finding was that the striatum was coding for the value of an outcome. A more global interpretation that takes into account all the neural inputs and outputs of this structure is that it takes in information about the outcome/reward, it matches up with the expectations (e.g., was the outcome better or worse than expected—if you guessed high was the card high, or did you make the wrong guess) and allows for the system to update and inform the next decision (e.g., maybe try low next time)."

17　computer guessed for them In response to a fact-checking email, Delgado expanded his comments: "There were three experiments related to this. . . . [In] the first one (Tricomi et al. 2004), they were told that they would see two circles. Upon seeing the yellow circle for example they would guess as before whether the correct answer was button 1 or 2 and were told that a correct response would yield a monetary reward. If they saw a blue circle they were told to press a button (motor control) but that the button had nothing to do with the reward, it was random. In truth, the reward was random in both cases, but if the subjects believed that their button press mattered, as in the yellow circle condition, then they engaged the striatum response much more than if it was a non-contingent reward. This experiment showed that if participants felt they were in control that the reward response was more prominent. The second experiment took this back to the card guessing game (Delgado et al. 2005) and this time added a cue, like a circle, before each trial that predicted if the card would be high or low. Participants had to learn via trial and error what the cue predicted. This experiment showed that the signal in the striatum was related to learning about the reward, rather than just purely processing the reward value. . . . In [the] third experiment (Leotti and Delgado 2005) we presented subjects with let's say two cues—a square and a circle. When they saw the square, they knew they would be faced with a 50/50 choice (a guess of sorts) and if they chose correctly, they would get a reward (no losses in this experiment, either a reward or no reward). In this condition, they felt in 'control.' Much like my participant who felt they could 'beat the game.' The other condition was the no-choice condition. Here, they saw a circle and were faced with the same choice. Except this time the computer picked for them. And if the computer was right they got a reward. So in both conditions one could get a reward (or no reward). But the key difference was that participants either had a choice or the computer chose. Interestingly, people preferred the choice condition, even though such condition required more effort (the actual choice) and led to the same amount of rewards. We also saw that the striatum activity was present to the square (compared to the circle). That is, when participants found out they had a choice, we saw activity in this reward area of the brain, suggesting that the mere opportunity for exerting one's choice may be rewarding in and of itself."

18 believed they were in control For more on Delgado's work, I recommend Elizabeth M. Tricomi, Mauricio R. Delgado, and Julie A. Fiez, "Modulation of Caudate Activity by Action Contingency," *Neuron* 41, no. 2 (2004): 281–92; Mauricio R. Delgado, M. Meredith Gillis, and Elizabeth A. Phelps, "Regulating the Expectation of Reward via Cognitive Strategies," *Nature Neuroscience* 11, no. 8 (2008): 880–81; Laura N. Martin and Mauricio R. Delgado, "The Influence of Emotion Regulation on Decision-Making Under Risk," *Journal of Cognitive Neuroscience* 23, no. 9 (2011): 2569–81; Lauren A. Leotti and Mauricio R. Delgado, "The Value of Exercising Control over Monetary Gains and Losses," *Psychological Science* 25, no. 2 (2014): 596–604; Lauren A. Leotti and Mauricio R. Delgado, "The Inherent Reward of Choice," *Psychological Science* 22 (2011): 1310–18.

18 reported to a boss "Self-Employment in the United States," *Monthly Labor Review,* U.S. Bureau of Labor Statistics, September 2010, http://www .bls.gov/opub/mlr/2010/09/art2full.pdf.

18 otherwise transitory positions A 2006 study by the Government Accountability Office found that 31 percent of workers were in temporary positions.

18 allocate their energy Michelle Conlin et al., "The Disposable Worker," *Bloomberg Businessweek,* January 7, 2010.

19 "The need for control" Lauren A Leotti, Sheena S. Iyengar, and Kevin N. Ochsner, "Born to Choose: The Origins and Value of the Need for Control," *Trends in Cognitive Sciences* 14, no. 10 (2010): 457–63.

19 setbacks faster Diana I. Cordova and Mark R. Lepper, "Intrinsic Motivation and the Process of Learning: Beneficial Effects of Contextualization, Personalization, and Choice," *Journal of Educational Psychology* 88, no. 4 (1996): 715; Judith Rodin and Ellen J. Langer, "Long-Term Effects of a Control-Relevant Intervention with the Institutionalized Aged," *Journal of Personality and Social Psychology* 35, no. 12 (1977): 897; Rebecca A. Henry and Janet A. Sniezek, "Situational Factors Affecting Judgments of Future Performance," *Organiza-*

tional Behavior and Human Decision Processes 54, no. 1 (1993): 104–32; Romin W. Tafarodi, Alan B. Milne, and Alyson J. Smith. "The Confidence of Choice: Evidence for an Augmentation Effect on Self-Perceived Performance," *Personality and Social Psychology Bulletin* 25, no. 11 (1999): 1405–16; Jack W. Brehm, "Postdecision Changes in the Desirability of Alternatives," *The Journal of Abnormal and Social Psychology* 52, no. 3 (1956): 384; Leon Festinger, *A Theory of Cognitive Dissonance*, vol. 2 (Stanford, Calif.: Stanford University Press, 1962); Daryl J. Bem, "An Experimental Analysis of Self-Persuasion," *Journal of Experimental Social Psychology* 1, no. 3 (1965): 199–218; Louisa C. Egan, Laurie R. Santos, and Paul Bloom, "The Origins of Cognitive Dissonance: Evidence from Children and Monkeys," *Psychological Science* 18, no. 11 (2007): 978–83.

19   longer than their peers E. J. Langer and J. Rodin, "The Effects of Choice and Enhanced Personal Responsibility for the Aged: A Field Experiment in an Institutional Setting," *Journal of Personality and Social Psychology* 34, no. 2 (1976): 191–98.

19   food into their mouths Margaret W. Sullivan and Michael Lewis, "Contextual Determinants of Anger and Other Negative Expressions in Young Infants," *Developmental Psychology* 39, no. 4 (2003): 693.

19   freedom to choose Leotti and Delgado, "Inherent Reward of Choice."

19   *Psychological Science* in 2011 Ibid.

20   autonomy and self-determination Erika A. Patall, Harris Cooper, and Jorgianne Civey Robinson, "The Effects of Choice on Intrinsic Motivation and Related Outcomes: A Meta-Analysis of Research Findings," *Psychological Bulletin* 134, no. 2 (2008): 270; Deborah J. Stipek and John R. Weisz, "Perceived Personal Control and Academic Achievement," *Review of Educational Research* 51, no. 1 (1981): 101–37; Steven W. Abrahams, "Goal-Setting and Intrinsic Motivation: The Effects of Choice and Performance Frame-of-Reference" (PhD diss., Columbia University, 1989); Teresa M. Amabile and Judith Gitomer, "Children's Artistic Creativity Effects of Choice in Task Materials," *Personality and Social Psychology Bulletin* 10, no. 2 (1984): 209–15; D'Arcy A. Becker, "The Effects of Choice on Auditors' Intrinsic Motivation and Performance," *Behavioral Research in Accounting* 9 (1997); Dan Stuart Cohen, "The Effects of Task Choice, Monetary, and Verbal Reward on Intrinsic Motivation: A Closer Look at Deci's Cognitive Evaluation Theory" (PhD diss., Ohio State University, 1974); Diana I. Cordova and Mark R. Lepper, "Intrinsic Motivation and the Process of Learning: Beneficial Effects of Contextualization, Personalization, and Choice," *Journal of Educational Psychology* 88, no. 4 (1996): 715; Hsiao d'Ailly, "The Role of Choice in Children's Learning: A Distinctive Cultural and Gender Difference in Efficacy, Interest, and Effort," *Canadian Journal of Behavioural Science* 36, no. 1 (2004): 17; Edward L. Deci, *The Psychology of Self-*

Determination (New York: Free Press, 1980); J. B. Detweiler, R. J. Mendoza, and M. R. Lepper, "Perceived Versus Actual Choice: High Perceived Choice Enhances Children's Task Engagement," 8th Annual Meeting of the American Psychological Society, San Francisco, 1996; John J. M. Dwyer, "Effect of Perceived Choice of Music on Exercise Intrinsic Motivation," *Health Values: The Journal of Health Behavior, Education and Promotion* 19, no. 2 (1995): 18–26; Gregory G. Feehan and Michael E. Enzle, "Subjective Control over Rewards: Effects of Perceived Choice of Reward Schedule on Intrinsic Motivation and Behavior Maintenance," *Perceptual and Motor Skills* 72, no. 3 (1991): 995–1006; Terri Flowerday, Gregory Schraw, and Joseph Stevens, "The Role of Choice and Interest in Reader Engagement," *The Journal of Experimental Education* 72, no. 2 (2004): 93–114; Claus A. Hallschmidt, "Intrinsic Motivation: The Effects of Task Choice, Reward Magnitude and Reward Choice" (PhD diss., University of Alberta, 1977); Sheena S. Iyengar and Mark R. Lepper, "Rethinking the Value of Choice: A Cultural Perspective on Intrinsic Motivation," *Journal of Personality and Social Psychology* 76, no. 3 (1999): 349; Keven A. Prusak et al., "The Effects of Choice on the Motivation of Adolescent Girls in Physical Education," *Journal of Teaching in Physical Education* 23, no. 1 (2004): 19–29; Johnmarshall Reeve, Glen Nix, and Diane Hamm, "Testing Models of the Experience of Self-Determination in Intrinsic Motivation and the Conundrum of Choice," *Journal of Educational Psychology* 95, no. 2 (2003): 375; Romin W. Tafarodi, Alan B. Milne, and Alyson J. Smith, "The Confidence of Choice: Evidence for an Augmentation Effect on Self-Perceived Performance," *Personality and Social Psychology Bulletin* 25, no. 11 (1999): 1405–16; Miron Zuckerman et al., "On the Importance of Self-Determination for Intrinsically-Motivated Behavior," *Personality and Social Psychology Bulletin* 4, no. 3 (1978): 443–46.

22   on a new life In response to a fact-checking email, Colonel Robert Gruny, commanding officer, Recruit Training Regiment, MCRD San Diego, wrote: "From the moment the recruits first step off of the bus onto the yellow footprints they are exposed to a degree of collective shock and stress that is designed to emphasize teamwork [and] obedience to orders and to reinforce the fact that they are entering into a new phase of their life in which selfless dedication to each other is coveted far more than individual achievement. In addition to the medical processing and haircuts reference above night one includes being checked for contraband, the very practical tasks of administrative processing and clothing issue, and making an initial call home to inform their parents or other designated individual that they have arrived safely at the Recruit Depot."

22   "their whole life" In response to a fact-checking email, Colonel Gruny wrote regarding Krulak's reforms: "The series of reforms were centered on the institution of values based training into recruit training and the introduction of the Crucible. While self-motivation and leadership were certainly enhanced by these reforms they also focused on teamwork, followership, and core values development (honor, courage, and commitment). Gen. Krulak sought to embed a training philosophy that resulted in our Marines making the right kind of values based decisions, in combat or in peacetime."

23 best course of action For my understanding of USMC boot camp, I am indebted to General Krulak and Major Neil A. Ruggiero, director of public affairs at MCRD San Diego/Western Recruiting Region. Additionally, I am indebted to Thomas E. Ricks and his book *Making the Corps* (New York: Scribner, 2007). I have also drawn upon Vincent Martino, Jason A. Santamaria, and Eric K. Clemons, *The Marine Corps Way: Using Maneuver Warfare to Lead a Winning Organization* (New York: McGraw-Hill, 2005); James Woulfe, *Into the Crucible: Making Marines for the 21st Century* (Novato, Calif.: Presidio Press, 2009); Jon R. Katzenbach, *Peak Performance: Aligning the Hearts and Minds of Your Employees* (Boston: Harvard Business Press, 2000); Megan M. Thompson and Donald R. McCreary, *Enhancing Mental Readiness in Military Personnel* (Toronto: Defense Research and Development, 2006); Ross R. Vickers Jr. and Terry L. Conway, "Changes in Perceived Locus of Control During Basic Training" (1984); Raymond W. Novaco et al., *Psychological and Organizational Factors Related to Attrition and Performance in Marine Corps Recruit Training*, no. AR-001 (Seattle: Washington University Department of Psychology, 1979); Thomas M. Cook, Raymond W. Novaco, and Irwin G. Sarason, "Military Recruit Training as an Environmental Context Affecting Expectancies for Control of Reinforcement," *Cognitive Therapy and Research* 6, no. 4 (1982): 409–27.

23 since the 1950s Julian B. Rotter, "Generalized Expectancies for Internal Versus External Control of Reinforcement," *Psychological Monographs: General and Applied* 80, no. 1 (1966): 1; Timothy A. Judge et al., "Are Measures of Self-Esteem, Neuroticism, Locus of Control, and Generalized Self-Efficacy Indicators of a Common Core Construct?" *Journal of Personality and Social Psychology* 83, no. 3 (2002): 693; Herbert M. Lefcourt, *Locus of Control: Current Trends in Theory and Research* (Hillsdale, N.J.: L. Erlbaum, 1982); Cassandra Bolyard Whyte, "High-Risk College Freshmen and Locus of Control," *Humanist Educator* 16, no. 1 (1977): 2–5; Angela Roddenberry and Kimberly Renk, "Locus of Control and Self-Efficacy: Potential Mediators of Stress, Illness, and Utilization of Health Services in College Students," *Child Psychiatry and Human Development* 41, no. 4 (2010): 353–70; Victor A. Benassi, Paul D. Sweeney, and Charles L. Dufour, "Is There a Relation Between Locus of Control Orientation and Depression?" *Journal of Abnormal Psychology* 97, no. 3 (1988): 357.

23 "Internal locus of control" Alexandra Stocks, Kurt A. April, and Nandani Lynton, "Locus of Control and Subjective Well-Being: A Cross-Cultural Study," *Problems and Perspectives in Management* 10, no. 1 (2012): 17–25.

24 difficult puzzles Claudia M. Mueller and Carol S. Dweck, "Praise for Intelligence Can Undermine Children's Motivation and Performance," *Journal of Personality and Social Psychology* 75, no 1 (1998): 33.

25 that study, told me The specific experiment conducted by Professor Dweck described in this chapter was focused on her implicit theory of intelligence rather than locus of control. In an interview, she drew comparisons between that work and its implications for understanding locus of control.

25 "in control of themselves" For more on Professor Dweck's fascinating research, I recommend Carol S. Dweck and Ellen L. Leggett, "A Social-Cognitive Approach to Motivation and Personality," *Psychological Review* 95, no. 2 (1988): 256; Carol S. Dweck, "Motivational Processes Affecting Learning," *American Psychologist* 41, no 10 (1986): 1040; Carol S. Dweck, Chi-yue Chiu, and Ying-yi Hong, "Implicit Theories and Their Role in Judgments and Reactions: A Word from Two Perspectives," *Psychological Inquiry* 6, no. 4 (1995): 267–85; Carol Dweck, *Mindset: The New Psychology of Success* (New York: Random House, 2006).

26 ketchup bottles In response to a fact-checking email, Colonel Jim Gruny, commanding officer, Recruit Training Regiment, MCRD San Diego, wrote that "this sounds like a scenario that may have been accurate at the time the Marine describing it experienced recruit training. Recruits no longer clean mess halls. That said, this scenario does accurately illustrate the methods used by our drill instructors and the lessons they seek to impart on our recruits."

27 obstacle courses In response to a fact-checking email, a spokesman for the USMC stressed that recruits are under supervision during the entirety of the Crucible, and that the area where the Crucible takes place is USMC property. In California, the Crucible takes place within Camp Pendleton; in Parris Island, South Carolina, it is an area around an old airstrip. Colonel Jim Gruny, commanding officer, Recruit Training Regiment, MCRD San Diego, wrote that "General Krulak pioneered the use of values-based training and a crucible to cement it among recruits. Krulak said his original intent for the Crucible as a culminating event was threefold. First, it would be the drill instructor's last opportunity to give a 'go or no go' to the individual recruit. Second, it would 'emphasize and reinforce all the core values training that was ongoing throughout recruit training' . . . Last, it would 'bring the recruit from an emphasis on self-discipline to where we want them to be in combat, which is selflessness.' . . . Failure to complete the Crucible may require a recruit to be recycled to another Company with which he can undergo the Crucible

again. He will only be dropped from the Marine Corps if he repeatedly fails to complete the Crucible or if he suffers an injury that prevents further military service." Colonel Christopher Nash, commanding officer, Weapons and Field Training Battalion, wrote: "The Crucible is a 54-hour endurance event that marks the transformation from civilian to U.S. Marine. Recruits, over a three day period, will travel approximately 68 km on foot, eat no more than three MREs for the duration of the event and operate with less than four hours of sleep a night. The focus of the Crucible is core values and teamwork. Recruits must overcome 24 stations/obstacles, participate in three core values discussions and two night endurance events during the three days. No event can be completed alone. The Crucible culminates with a 16 km 'Reaper' hike in which an emblem ceremony occurs. During this event recruits earn the title Marine."

30 during basic training Joey E. Klinger, "Analysis of the Perceptions of Training Effectiveness of the Crucible at Marine Corps Recruit Depot, San Diego" (PhD diss., Naval Postgraduate School, 1999); S. P. Dynan, *Updating Tradition: Necessary Changes to Marine Corps Recruit Training* (Quantico, Va.: Marine Corps Command and Staff College, 2006); M. C. Cameron, *Crucible Marine on Point: Today's Entry-Level Infantry Marine* (Quantico, Va.: Marine Corps Command and Staff College, 2006); Michael D. Becker, "'We Make Marines': Organizational Socialization and the Effects of 'The Crucible' on the Values Orientation of Recruits During US Marine Corps Training" (PhD diss., Indiana University of Pennsylvania, 2013); Benjamin Eiseman, "Into the Crucible: Making Marines for the 21st Century," *Military Review* 80, no. 1 (2000): 94; Terry Terriff, "Warriors and Innovators: Military Change and Organizational Culture in the US Marine Corps," *Defense Studies* 6, no. 2 (2006): 215–47; Antonio B. Smith, *United States Marine Corps' Entry-Level Training for Enlisted Infantrymen: The Marginalization of Basic Warriors* (Quantico, Va.: Marine Corps Command and Staff College, 2001); William Berris, *Why General Krulak Is the Marine Corps' Greatest Strategic Leader* (Carlisle Barracks, Penn.: U.S. Army War College, 2011); Terry Terriff, "Of Romans and Dragons: Preparing the US Marine Corps for Future Warfare," *Contemporary Security Policy* 28, no. 1 (2007): 143–62; Marie B. Caulfield, *Adaptation to First Term Enlistment Among Women in the Marine Corps* (Boston: Veterans Administration Medical Center, 2000); Craig M. Kilhenny, "An Organizational Analysis of Marine Corps Recruit Depot, San Diego" (PhD diss., Naval Postgraduate School, 2003); Larry Smith, *The Few and the Proud: Marine Corps Drill Instructors in Their Own Words* (New York: W. W. Norton, 2007); Thomas M. Cook, Raymond W. Novaco, and Irwin G. Sarason, "Military Recruit Training as an Environmental Context Affecting Expectancies for Control of Reinforcement," *Cognitive Therapy and Research* 6, no. 4 (1982): 409–27; Ross R. Vickers Jr.

and Terry L. Conway, *The Marine Corps Basic Training Experience: Psychosocial Predictors of Performance, Health, and Attrition* (San Diego: Naval Health Research Center, 1983); Ross R. Vickers Jr. and Terry L. Conway, "Changes in Perceived Locus of Control During Basic Training" (paper presented at the Annual Meeting of the American Psychological Association: Toronto, Canada, August 24–28 (1984); Thomas M. Cook, Raymond W. Novaco, and Irwin G. Sarason, *Generalized Expectancies, Life Experiences, and Adaptation to Marine Corps Recruit Training* (Seattle: Washington University: Department of Psychology, 1980); R. R. Vickers Jr. et al., *The Marine Corps Training Experience: Correlates of Platoon Attrition Rate Differences* (San Diego: Naval Health Research Center, 1983).

31 force upon them Rosalie A. Kane et al., "Everyday Matters in the Lives of Nursing Home Residents: Wish for and Perception of Choice and Control," *Journal of the American Geriatrics Society* 45, no. 9 (1997): 1086–93; Rosalie A. Kane et al., "Quality of Life Measures for Nursing Home Residents," *The Journals of Gerontology Series A: Biological Sciences and Medical Sciences* 58, no. 3 (2003): 240–48; James R. Reinardy and Rosalie A. Kane, "Anatomy of a Choice: Deciding on Assisted Living or Nursing Home Care in Oregon," *Journal of Applied Gerontology* 22, no. 1 (2003): 152–74; Robert L. Kane and Rosalie A. Kane, "What Older People Want from Long-Term Care, and How They Can Get It," *Health Affairs* 20, no. 6 (2001): 114–27; William J. McAuley and Rosemary Blieszner, "Selection of Long-Term Care Arrangements by Older Community Residents," *The Gerontologist* 25, no. 2 (1985): 188–93; Bart J. Collopy, "Autonomy in Long Term Care: Some Crucial Distinctions," *The Gerontologist* 28, supplement (1988): 10–17; Elizabeth H. Bradley et al., "Expanding the Andersen Model: The Role of Psychosocial Factors in Long-Term Care Use," *Health Services Research* 37, no. 5 (2002): 1221–42; Virginia G. Kasser and Richard M. Ryan, "The Relation of Psychological Needs for Autonomy and Relatedness to Vitality, Well-Being, and Mortality in a Nursing Home: Effects of Control and Predictability on the Physical and Psychological Well-Being of the Institutionalized Aged," *Journal of Applied Social Psychology* 29, no. 5 (1999): 935–54; James F. Fries, "The Compression of Morbidity," *The Milbank Memorial Fund Quarterly: Health and Society* 83, no. 4 (2005): 801–23; Richard Schulz, "Effects of Control and Predictability on the Physical and Psychological Well-Being of the Institutionalized Aged," *Journal of Personality and Social Psychology* 33, no. 5 (1976): 563.

35 They didn't feel anything In response to a fact-checking email, Habib expanded upon his comments and said that rather than categorize the patients as not understanding feelings, it might be more accurate to say "it is a matter of expression of feelings, more than feeling itself. They can recall what they felt before, and there is no evidence they cannot feel it anymore. Instead, it seems that since they have no more manifestations of seeking satisfaction, they look like they had no feeling. This is also an intriguing observation, since it suggests that the intensity of feelings is dependent upon the individual's capacity of seeking satisfaction or reward."

CHAPTER TWO: TEAMS

39 school's websites explained Alex Roberts, "What a Real Study Group Looks Like," Yale School of Management, *MBA Blog*, August 31, 2010, http://som.yale.edu/what-real-study-group-looks.

40 "didn't gel." In an email sent in response to fact-checking questions, Julia Rozovsky wrote: "There were a few members of my study group that I developed close friendships with, however I was much closer to my case study team."

41 first in the nation "Yale SOM Team Wins National Net Impact Case Competition," Yale School of Management, November 10, 2011, http://som.yale.edu/news/news/yale-som-team-wins-national-net-impact-case-competition.

41 were at Yale In an email sent in response to fact-checking questions, Julia Rozovsky wrote: "We chose to enter the competition each time. Each competition was a separate team/entry/packet/process. I just happened to work with the same team fairly consistently."

42 spent their time In an email sent in response to fact-checking questions, a Google spokeswoman wrote that "People Analytics' overarching theme is that we study the key drivers of Health, Happiness and Productivity of Googlers in a scientific and rigorous way. . . . No one part of Google controls or oversees hiring or promo, but rather it is shared with Googlers themselves, with managers, etc." For more on Google's approach to human resources, please see Thomas H. Davenport, Jeanne Harris, and Jeremy Shapiro, "Competing on Talent Analytics," *Harvard Business Review* 88, no. 10 (2010): 52–58; John Sullivan, "How Google Became the #3 Most Valuable Firm by Using People Analytics to Reinvent HR," ERE Media, February 25, 2013, http://www.eremedia.com/ere/how-google-became-the-3-most-valuable-firm-by-using-people-analytics-to-reinvent-hr/; David A. Garvin, "How Google Sold Its Engineers on Management," *Harvard Business Review* 91, no. 12 (2013): 74–82; Adam Bryant, "Google's Quest to Build a Better Boss," *The New York Times*, March 12, 2011; Laszlo Bock, *Work Rules! Insights from Inside Google That Will Transform the Way You Live and Lead* (New York: Twelve, 2015).

42 America's top workplaces In 2007, 2008, 2012, 2013, and 2014, Google was ranked number one by *Fortune*.

43   one of the effort's researchers In an email sent in response to fact-checking questions, Julia Rozovsky wrote: "I worked on several other efforts prior to joining the Project Aristotle team. Here's a quick bio that I use internally: 'Julia Rozovsky joined Google's People Analytics team in August 2012. During her time at Google, Julia has advised teams on workforce planning and design strategies, analyzed the impact of workplace flexibility programs, and conducted research on empowering leaders. She is currently the [project manager] of Project Aristotle, which aims to improve team effectiveness at Google. Prior to Google, Julia collaborated with Harvard Business School academics on competitive strategy and organizational behavior research focusing specifically on game theory, ethics and financial controls, and organizational structure. Earlier in her career, Julia was a strategy consultant with a boutique marketing analytics firm. Julia holds an MBA from the Yale School of Management, and a BA in mathematics and economics from Tufts University.'"

44   made a team effective In comments sent in response to fact-checking questions, a Google spokeswoman wrote: "The first thing we had to start with was the definition of a team, and we arrived at groups of people collaborating closely on projects and working toward a common goal. Then, since we knew a hierarchical team definition would be too limiting in our environment where people collaborate across reporting lines, we had to figure out how to systematically identify intact teams and their accurate membership so we could study them. In the end, we had to do it manually, by asking senior leaders to identify teams in their orgs and ask the teams' leads to confirm the members."

45   "appropriate behavior" David Lyle Light Shields et al., "Leadership, Cohesion, and Team Norms Regarding Cheating and Aggression," *Sociology of Sport Journal* 12 (1995): 324–36.

45   deference to the team For more on norms, please see Muzafer Sherif, *The Psychology of Social Norms* (London: Octagon Books, 1965); Jay Jackson, "Structural Characteristics of Norms," *Current Studies in Social Psychology* 301 (1965): 309; P. Wesley Schultz et al., "The Constructive, Destructive, and Reconstructive Power of Social Norms," *Psychological Science* 18, no. 5 (2007): 429–34; Robert B. Cialdini, "Descriptive Social Norms as Underappreciated Sources of Social Control," *Psychometrika* 72, no. 2 (2007): 263–68; Keithia L. Wilson et al., "Social Rules for Managing Attempted Interpersonal Domination in the Workplace: Influence of Status and Gender," *Sex Roles* 44, nos. 3–4 (2001): 129–54; Daniel C. Feldman, "The Development and Enforcement of Group Norms," *Academy of Management Review* 9, no. 1 (1984): 47–53; Deborah J. Terry, Michael A. Hogg, and Katherine M. White, "The Theory of Planned Behaviour: Self-Identity, Social Identity and Group Norms," *The British Journal of Social Psychology* 38 (1999): 225; Jolanda Jetten, Russell Spears, and Antony S. R. Manstead, "Strength of Identification and Intergroup Differentiation: The Influence of Group Norms," *European Journal of Social Psychology* 27, no. 5 (1997): 603–9; Mark G. Ehrhart and Stefanie E. Naumann, "Organizational Citizenship Behavior in Work Groups: A Group Norms Approach," *Journal of Applied Psychology* 89, no. 6 (2004): 960; Daniel C. Feldman, "The Development and Enforcement of Group Norms," *Academy of Management Review* 9, no. 1 (1984): 47–53; Jennifer A. Chatman and Francis J. Flynn, "The Influence of Demographic Heterogeneity on the Emergence and Consequences of Cooperative Norms in Work Teams," *Academy of Management Journal* 44, no. 5 (2001): 956–74.

46 discouraged by our teammates Sigal G. Barsade, "The Ripple Effect: Emotional Contagion and Its Influence on Group Behavior," *Administrative Science Quarterly* 47, no. 4 (2002): 644–75; Vanessa Urch Druskat and Steven B. Wolff, "Building the Emotional Intelligence of Groups," *Harvard Business Review* 79, no. 3 (2001): 80–91; Vanessa Urch Druskat and Steven B. Wolff, "Group Emotional Intelligence and Its Influence on Group Effectiveness," in *The Emotionally Intelligent Workplace: How to Select for, Measure, and Improve Emotional Intelligence in Individuals, Groups and Organizations,* ed. Cary Cherniss and Daniel Goleman (San Francisco: Jossey-Bass, 2001), 132–55; Daniel Goleman, Richard Boyatzis, and Annie McKee, "The Emotional Reality of Teams," *Journal of Organizational Excellence* 21, no. 2 (2002): 55–65; William A. Kahn, "Psychological Conditions of Personal Engagement and Disengagement at Work," *Academy of Management Journal* 33, no. 4 (1990): 692–724; Tom Postmes, Russell Spears, and Sezgin Cihangir, "Quality of Decision Making and Group Norms," *Journal of Personality and Social Psychology* 80, no. 6 (2001): 918; Chris Argyris, "The Incompleteness of Social-Psychological Theory: Examples from Small Group, Cognitive Consistency, and Attribution Research," *American Psychologist* 24, no. 10 (1969): 893; James R. Larson and Caryn Christensen, "Groups as Problem-Solving Units: Toward a New Meaning of Social Cognition," *British Journal of Social Psychology* 32, no. 1 (1993): 5–30; P. Wesley Schultz et al., "The Constructive, Destructive, and Reconstructive Power of Social Norms," *Psychological Science* 18, no. 5 (2007): 429–34.

46 put her on guard In an email sent in response to fact-checking questions, Julia Rozovsky wrote: "This is how the study group felt from time to time. Not consistently."

46 equally successful group In comments sent in response to fact-checking questions, a Google spokeswoman wrote: "We wanted to test many group norms that we thought might be important. But at the testing phase we didn't know that the *how* was going to be more important than the *who*. When we started running the statistical models, it became clear that not only were the norms more important in our models but that 5 themes stood out from the rest."

47 **Boston hospitals** Amy C. Edmondson, "Learning from Mistakes Is Easier Said than Done: Group and Organizational Influences on the Detection and Correction of Human Error," *The Journal of Applied Behavioral Science* 32, no. 1 (1996): 5–28; Druskat and Wolff, "Group Emotional Intelligence," 132–55; David W. Bates et al., "Incidence of Adverse Drug Events and Potential Adverse Drug Events: Implications for Prevention," *Journal of the American Medical Association* 274, no. 1 (1995): 29–34; Lucian L. Leape et al., "Systems Analysis of Adverse Drug Events," *Journal of the American Medical Association* 274, no. 1 (1995): 35–43.

47 **"slip through the cracks"** In an email sent in response to fact-checking questions, Edmondson wrote: "It's not MY insight that mistakes occur because of system complexity (and its challenging combination with patient heterogeneity). . . . I am merely the messenger bringing that perspective to certain audiences. But yes, the opportunities for slipping through are ever-present, so the challenge is building awareness and teamwork that catch and correct and prevent the slips."

49 **teammates behaved** In an email sent in response to fact-checking questions, Edmondson wrote: "My goal was to figure out whether the interpersonal climate that I'd found to differ in this setting would differ in other organizations, especially in terms of differing between groups within the same organization. Later I called this psychological safety (or team psychological safety). I also wanted to discover whether, if it did differ, whether that difference would be associated with differences in learning behavior (and in performance)." For more on Edmondson's work, please see Amy C. Edmondson, "Psychological Safety and Learning Behavior in Work Teams," *Administrative Science Quarterly* 44, no. 2 (1999): 350–83; Ingrid M. Nembhard and Amy C. Edmondson, "Making It Safe: The Effects of Leader Inclusiveness and Professional Status on Psychological Safety and Improvement Efforts in Health Care Teams," *Journal of Organizational Behavior* 27, no. 7 (2006): 941–66; Amy C. Edmondson, Roderick M. Kramer, and Karen S. Cook, "Psychological Safety, Trust, and Learning in Organizations: A Group-Level Lens," *Trust and Distrust in Organizations: Dilemmas and Approaches* 10 (2004): 239–72; Amy C. Edmondson, *Managing the Risk of Learning: Psychological Safety in Work Teams* (Boston: Division of Research, Harvard Business School, 2002); Amy C. Edmondson, Richard M. Bohmer, and Gary P. Pisano, "Disrupted Routines: Team Learning and New Technology Implementation in Hospitals," *Administrative Science Quarterly* 46, no. 4 (2001): 685–716; Anita L. Tucker and Amy C. Edmondson, "Why Hospitals Don't Learn from Failures," *California Management Review* 45, no. 2 (2003): 55–72; Amy C. Edmondson, "The Competitive Imperative of Learning," *Harvard Business Review* 86, nos. 7–8 (2008): 60; Amy C. Edmondson, "A Safe Harbor: Social Psychological Conditions Enabling Bound-

ary Spanning in Work Teams," Research on Managing Groups and Teams 2 (1999): 179–99; Amy C. Edmondson and Kathryn S. Roloff, "Overcoming Barriers to Collaboration: Psychological Safety and Learning in Diverse Teams," *Team Effectiveness in Complex Organizations: Cross-Disciplinary Perspectives and Approaches* 34 (2009): 183–208.

50 1999 paper Amy C. Edmondson, "Psychological Safety and Learning Behavior in Work Teams," *Administrative Science Quarterly* 44, no. 2 (1999): 350–83.

50 her Google colleagues In an email responding to fact-checking questions, a Google spokeswoman wrote: "We found Edmondson's papers on psych safety very useful when trying to figure out how to cluster norms that we saw popping up as important into meta-themes. When we reviewed the papers about psych safety, we noticed that norms like allowing others to fail without repercussions, respecting divergent opinions, feeling as if others aren't trying to undermine you are all part of psychological safety. This became one of our five key themes, along with dependability, structure/clarity, job meaning, and impact."

51 would never stop For my understanding of the early days of *Saturday Night Live,* I am indebted to those writers and cast members who were willing to speak with me, as well as Tom Shales and James Andrew Miller, *Live from New York: An Uncensored History of "Saturday Night Live"* (Boston: Back Bay Books, 2008); Ellin Stein, *That's Not Funny, That's Sick: The National Lampoon and the Comedy Insurgents Who Captured the Mainstream* (New York: Norton, 2013); Marianne Partridge, ed., *"Rolling Stone" Visits "Saturday Night Live"* (Garden City, N.Y.: Dolphin Books, 1979); Doug Hill and Jeff Weingrad, *Saturday Night: A Backstage History of "Saturday Night Live"* (San Francisco: Untreed Reads, 2011).

53 "never be heard from again" In an email sent in response to a fact-checking question, Schiller wrote: "It was an intense experience for me since I had never lived in New York or worked on a comedy-variety show. A lot of us were new to Manhattan and as such, hung out a lot together not only because New York at that time was sort of dangerous and scary, but also we didn't know that many people and we were formulating the show. We were in our midtwenties and early thirties. Yes, we'd eat at restaurants and go to bars together even when out of the studio. We moved en masse, trying to make each other laugh."

54 "among the show's cast" Malcolm Gladwell, "Group Think: What Does *Saturday Night Live* Have in Common with German Philosophy?" *The New Yorker,* December 2, 2002.

54 team intensely bonds Donelson Forsyth, *Group Dynamics* (Boston: Cengage Learning, 2009).

54 "It was a stalag" Alison Castle, *"Saturday Night Live": The Book* (Reprint, Cologne: Taschen, America, 2015).

55 "someone else was failing" In an email sent in response to a fact-checking question, Beatts wrote: "My Holocaust joke, which was certainly said in jest because there is no other way to say a joke, had nothing whatsoever to do with the show's writers. The exact wording was 'Imagine if Hitler hadn't killed six million Jews, how hard it would be to find an apartment in New York.' It was a joke about the difficulty of finding apartments in New York, riffing off New York's large Jewish population and general ethnic feeling, a la 'You don't have to be Jewish to love Levy's rye bread. But it wouldn't hurt.' Zero to do with the writers. Marilyn Miller took offense at the mere mention of Hitler and the Holocaust, which to her could not be a subject for comedy. . . . [Regarding] competition among the writers, not that it didn't exist, because it did, but . . . everyone always had a chance to come back swinging the following week. Also the other writers and everyone in general, despite the competition for airtime, Lorne's approval, audience appreciation, etc., were always very supportive of other people's efforts and sympathetic to each other's failures. No one went around rubbing their hands in glee and going haha, your sketch was cut and mine wasn't, so there! It was more an attitude of 'Better luck next time.' I think everyone felt part of a family, maybe a dysfunctional family, but a close-knit family all the same. I would say that there is more backstabbing and jealousy and rivalry and competition and cliqueishness on the average middle school playground than there ever was at *SNL* during the time I was there."

55 "stuff for other people" In an email sent in response to fact-checking questions, Alan Zweibel wrote: "I wasn't angry because of anything to do with that character or the process in which it was written. She and I weren't speaking for reasons that I really can't recall. But after about three shows where I didn't write with her (and for her) we both realized that our work was suffering—that we were better as a team than we were individually—so we buried the hatchet and began collaborating again."

56 "it could be brutal" In an email sent in response to a fact-checking question, Schiller wrote: "I would say that some, not all, comedy writers and stand-up comedians have some sadness or anger in their life that helped fuel their comedy. They are fast with quips, and the stand-ups were used to hecklers and had to be prepared with a quick comeback. So just as much as they can say something sharply funny, they can also jab you with a quick, hostile (but also funny) remark. . . . The atmosphere at *SNL*, although we all liked each other, could become highly competitive based on the fact that there were 10 writers and only so many sketches could go on the show, so we all did our best to write the winning sketch or make (in my case) the best short film."

58   **58 percent** The correct answers to the quiz are upset, decisive, skeptical, and cautious. These images come from Simon Baron-Cohen et al., "Another Advanced Test of Theory of Mind: Evidence from Very High Functioning Adults with Autism or Asperger Syndrome," *Journal of Child Psychology and Psychiatry* 38, no. 7 (1997): 813–22. And Simon Baron-Cohen et al., "The 'Reading the Mind in the Eyes' Test Revised Version: A Study with Normal Adults, and Adults with Asperger Syndrome or High-Functioning Autism," *Journal of Child Psychology and Psychiatry* 42, no. 2 (2001): 241–51.

58   *Science* **in 2010** Anita Williams Woolley et al., "Evidence for a Collective Intelligence Factor in the Performance of Human Groups," *Science* 330, no. 6004 (2010): 686–88.

60   **"individuals in it"** Anita Woolley and Thomas Malone, "What Makes a Team Smarter? More Women," *Harvard Business Review* 89, no. 6 (2011): 32–33; Julia B. Bear and Anita Williams Woolley, "The Role of Gender in Team Collaboration and Performance," *Interdisciplinary Science Reviews* 36, no. 2 (2011): 146–53; David Engel et al., "Reading the Mind in the Eyes or Reading Between the Lines? Theory of Mind Predicts Collective Intelligence Equally Well Online and Face-to-Face," *PloS One* 9, no. 12 (2014); Anita Williams Woolley and Nada Hashmi, "Cultivating Collective Intelligence in Online Groups," in *Handbook of Human Computation,* ed. Pietro Michelucci (New York: Springer, 2013), 703–14; Heather M. Caruso and Anita Williams Woolley, "Harnessing the Power of Emergent Interdependence to Promote Diverse Team Collaboration," *Research on Managing Groups and Teams: Diversity and Groups* 11 (2008): 245–66; Greg Miller, "Social Savvy Boosts the Collective Intelligence of Groups," *Science* 330, no. 6000 (2010): 22; Anita Williams Woolley et al., "Using Brain-Based Measures to Compose Teams: How Individual Capabilities and Team Collaboration Strategies Jointly Shape Performance," *Social Neuroscience* 2, no. 2 (2007): 96–105; Peter Gwynne, "Group Intelligence, Teamwork, and Productivity," *Research Technology Management* 55, no. 2 (2012): 7.

61   **University of Cambridge** Baron-Cohen et al., " 'Reading the Mind in the Eyes' Test Revised Version," 241–51.

63   **"more initials he sees"** In an email sent in response to fact-checking questions, Alan Zweibel wrote: "[Michaels] had said that he likes when there's a lot of initials at the top of the page because it meant that it had a variety of input and sensibilities. I believe that the show has lasted 40 years because Lorne is a genius when it comes to recognizing talent, rolling with the changing times, and encouraging everyone (while developing their individual voices) to work with each other so the total is greater than the sum of its parts."

63   **"the pain!"**   In the script that made it to air, O'Donoghue says, "'I know I can! I know I can! I know I can! I know I can! Heart attack! Heart attack! Heart attack! Heart attack! Oh, my God, the pain! Oh, my God, the pain! Oh, my God, the pain!" It is worth noting that the original concept for depressing children stories originated with O'Donoghue, not Garrett.

CHAPTER THREE: FOCUS

72   **bound for Paris**   For my understanding of the details of Air France Flight 447, I am indebted to numerous experts, including William Langewiesche, Steve Casner, Christopher Wickens, and Mica Endsley. I also drew heavily on a number of publications: William Langewiesche, "The Human Factor," *Vanity Fair*, October 2014; Nicola Clark, "Report Cites Cockpit Confusion in Air France Crash," *The New York Times*, July 6, 2012; Nicola Clark, "Experts Say Pilots Need More Air Crisis Training," *The New York Times*, November 21, 2011; Kim Willsher, "Transcripts Detail the Final Moments of Flight from Rio," *Los Angeles Times*, October 16, 2011; Nick Ross and Neil Tweedie, "Air France Flight 447: 'Damn It, We're Going to Crash,'" *The Daily Telegraph*, May 1, 2012; "Air France Flight 447: When All Else Fails, You Still Have to Fly the Airplane," *Aviation Safety*, March 1, 2011; "Concerns over Recovering AF447 Recorders," *Aviation Week*, June 3, 2009; Flight Crew Operating Manual, *Airbus 330—Systems—Maintenance System;* Tim Vasquez, "Air France Flight 447: A Detailed Meteorological Analysis," Weather Graphics, June 3, 2009, http://www.weathergraphics.com/tim/af447/; Cooperative Institute for Meteorological Satellite Studies, "Air France Flight #447: Did Weather Play a Role in the Accident?" *CIMSS Satellite Blog*, June 1, 2009, http://cimss.ssec.wisc.edu/goes/blog/archives/2601; Richard Woods and Matthew Campbell, "Air France 447: The Computer Crash," *The Times*, June 7, 2009; "AF 447 May Have Come Apart Before Crash," Associated Press, June 3, 2009; Wil S. Hylton, "What Happened to Air France Flight 447?" *The New York Times Magazine*, May 4, 2011; "Accident Description F-GZC," Flight Safety Foundation, Web; "List of Passengers Aboard Lost Air France Flight," Associated Press, June 4, 2009; "Air France Jet 'Did Not Break Up in Mid-Air,' Air France Crash: First Official Airbus A330 Report Due by Air Investigations and Analysis Office," *Sky News*, July 2, 2009; Matthew Wald, "Clues Point to Speed Issues in Air France Crash," *The New York Times*, June 7, 2009; Air France, "AF 447 RIO-PARIS-CDG, Pitot Probes," October 22, 2011, http://corporate.airfrance.com/en/press/af-447-rio-paris-cdg/pitot-probes/; Edward Cody, "Airbus Recommends Airlines Replace Speed Sensors," *The Washington Post,* July 31, 2009; Jeff Wise, "What Really Happened Aboard Air France 447," *Popular Mechanics,* December 6, 2011; David Kaminski-Morrow, "AF447 Stalled but Crew

Maintained Nose-Up Attitude," *Flight International*, May 27, 2011; David Talbot, "Flight 447's Fatal Attitude Problem," *Technology Review*, May 27, 2011; Glenn Pew, "Air France 447—How Did This Happen?" *AVweb*, May 27, 2011; Bethany Whitfield, "Air France 447 Stalled at High Altitude, Official BEA Report Confirms," *Flying*, May 27, 2011; Peter Garrison, "Air France 447: Was It a Deep Stall?" *Flying*, June 1, 2011; Gerald Traufetter, "Death in the Atlantic: The Last Four Minutes of Air France Flight 447," *Spiegel Online*, February 25, 2010; Nic Ross and Jeff Wise, "How Plane Crash Forensics Lead to Safer Aviation," *Popular Mechanics*, December 18, 2009; *Interim Report on the Accident on 1 June 2009 to the Airbus A330-203 Registered F-GZCP Operated by Air France Flight AF 447 Rio de Janeiro–Paris* (Paris: Bureau d'Enquêtes et d'Analyses pour la sécurité de l'aviation civile [BEA], 2012); *Interim Report No. 3 on the Accident on 1 June 2009 to the Airbus A330-203 registered F-GZCP Operated by Air France Flight AF 447 Rio de Janeiro–Paris* (Paris: BEA, 2011); *Final Report on the Accident on 1st June 2009 to the Airbus A330-203 Registered F-GZCP Operated by Air France Flight AF 447 Rio de Janeiro–Paris* (Paris: BEA, 2012); "Appendix 1 to *Final Report on the Accident on 1st June 2009 to the Airbus A330-203 Registered F-GZCP Operated by Air France Flight AF 447 Rio de Janeiro–Paris*" (Paris: BEA, July 2012); *Lost: The Mystery of Flight 447*, BBC One, June 2010; "Crash of Flight 447," *Nova*, 2010, produced by Nacressa Swan; "Air France 447, One Year Out," *Nova*, 2010, produced by Peter Tyson.

72 flying them home Air France has argued that it is inappropriate to blame pilot error as the primary cause for the crash of Flight 447. (This perspective is disputed by numerous aviation experts.) Air France was presented with a complete list of questions regarding details discussed in this chapter. The airline declined to comment on issues that fell outside of those topics discussed in the official report regarding Air France Flight 447 published by the Bureau d'Enquêtes et d'Analyses pour la sécurité de l'aviation civile, or BEA, which is the French authority responsible for investigating aviation accidents. In a statement, a spokesman for Air France wrote: "It is essential to remember that the BEA investigation report, the only official and public investigation to date, discusses and develops many of the subjects mentioned [in this chapter]. This report is available on the BEA website in English. We can only direct the journalist to this report to supplement our answers."

72 rotated responsibilities In response to questions, a spokesman for Air France noted that automation on long-haul aircraft preceded the A330 by twenty years, and that at one time "the crew included a flight engineer, who was responsible for monitoring all aircraft systems during the flight. On modern aircraft, the flight engineer has disappeared, but the requirement of monitoring aircraft systems remains. This is carried out by the pilots. Finally, now as in the past, beyond a certain flight time the crew is reinforced by one or more additional pilots to enable each pilot to take a rest period."

72   crashed after takeoff Isabel Wilkerson, "Crash Survivor's Psychic Pain May Be the Hardest to Heal," *The New York Times,* August 22, 1987; Mike Householder, "Survivor of 1987 Mich. Plane Crash Breaks Silence," Associated Press, May 15, 2013.

73   One hundred and one people Ninety-nine people were killed instantly in this crash. Two later died from complications.

73   into the Everglades Ken Kaye, "Flight 401 1972 Jumbo Jet Crash Was Worst Aviation Disaster in State History," *Sun Sentinel,* December 29, 1992.

73   other human errors Aviation Safety Network, NTSB records.

74   ascended by three thousand feet In response to questions, a spokesman for Air France wrote: "It has not been shown by the BEA that the action to pitch up is the result of the pilot's actions faced with the rolling of the aircraft, but rather the loss of altitude read, the vertical speed on descent of 600 ft per minute, the noise, the pitch that had diminished during the seconds before etc."

75   said Bonin In response to questions, a spokesman for Air France wrote: "What is written is true, but does not throw light comprehensively on this phase because of the lack of some essential elements, such as the fact that the STALL alarm went off twice at the beginning of the incident which may have led the pilots to doubt its validity when it went off repeatedly. The BEA report stated that audio alarms are not 'unmissable' and that on the contrary they are often the first to be ignored."

75   watching the kids Zheng Wang and John M. Tchernev, "The 'Myth' of Media Multitasking: Reciprocal Dynamics of Media Multitasking, Personal Needs, and Gratifications," *Journal of Communication* 62, no. 3 (2012): 493–513; Daniel T. Willingham, *Cognition: The Thinking Animal,* 3rd ed. (Upper Saddle River, N.J.: Pearson, 2007).

76   by automation Juergan Kiefer et al., "Cognitive Heuristics in Multitasking Performance," Center of Human-Machine Systems, Technische Universität Berlin, 2014, http://www.prometei.de/fileadmin/prometei.de/publikationen/Kiefer_eurocogsci2007.pdf.

76   automaticity and focus Barnaby Marsh et al., "Cognitive Heuristics: Reasoning the Fast and Frugal Way," in *The Nature of Reasoning,* eds. J. P. Leighton and R. J. Sternberg (New York: Cambridge University Press, 2004); "Human Performance," Aerostudents, http://aerostudents.com/files/human MachineSystems/humanPerformance.pdf.

76   misstep can be tragic For more on this topic, I particularly recommend Martin Sarter, Ben Givens, and John P. Bruno, "The Cognitive Neuroscience of Sustained Attention: Where Top-Down Meets Bottom-Up," *Brain Research Reviews* 35, no. 2 (2001): 146–60; Michael I. Posner and Steven E. Petersen, "The Attention System of the Human Brain," *Annual Review of Neuroscience* 13, no. 1 (1990): 25–42; Eric I. Knudsen, "Fundamental Components of Attention," *Annual Review of Neuroscience* 30 (2007): 57–78; Steven E. Petersen and Michael I. Posner, "The Attention System of the Human Brain: 20 Years After," *Annual Review of Neuroscience* 35 (2012): 73; Raja Parasuraman, Robert Molloy, and Indramani L. Singh, "Performance Consequences of Automation-Induced 'Complacency,'" *The International Journal of Aviation Psychology* 3, no. 1 (1993): 1–23; Raymond S. Nickerson et al., *Handbook of Applied Cognition*, ed. Francis T. Durso (Hoboken, N.J.: Wiley, 2007); Christopher D. Wickens, "Attention in Aviation," University of Illinois at Urbana-Champaign Institute of Aviation, Research Gate, February 1987, http://www.researchgate .net/publication/4683852_Attention_in_aviation; Christopher D. Wickens, "The Psychology of Aviation Surprise: An 8 Year Update Regarding the Noticing of Black Swans," *Proceedings of the 15th International Symposium on Aviation Psychology*, 2009.

76   critical than ever before   Ludwig Reinhold Geissler, "The Measurement of Attention," *The American Journal of Psychology* (1909): 473–529; William A. Johnston and Steven P. Heinz, "Flexibility and Capacity Demands of Attention," *Journal of Experimental Psychology: General* 107, no. 4 (1978): 420; Robin A. Barr, "How Do We Focus Our Attention?" *The American Journal of Psychology* (1981): 591–603.

76   panicked attention   G. R. Dirkin, "Cognitive Tunneling: Use of Visual Information Under Stress," *Perceptual and Motor Skills* 56, no. 1 (1983): 191–98; David C. Foyle, Susan R. Dowell, and Becky L. Hooey, "Cognitive Tunneling in Head-Up Display (HUD) Superimposed Symbology: Effects of Information Location" (2001); Adrien Mack and Irvin Rock, *Inattentional Blindness* (Cambridge, Mass.: MIT Press, 2000); Steven B. Most, Brian J. Scholl, Daniel J. Simons, and Erin R. Clifford, "What You See Is What You Get: Sustained Inattentional Blindness and the Capture of Awareness," *Psychological Review* 112, no. 1 (2005): 217–42; Daniel J. Simons, "Attentional Capture and Inattentional Blindness," *Trends in Cognitive Sciences* 4, no. 4 (2000): 147–55; Gustav Kuhn and Benjamin W. Tatler, "Misdirected by the Gap: The Relationship Between Inattentional Blindness and Attentional Misdirection," *Consciousness and Cognition* 20, no. 2 (2011): 432–36; William J. Horrey and Christopher D. Wickens, "Examining the Impact of Cell Phone Conversations on Driving Using Meta-Analytic Techniques," *Human Factors: The Journal of the Human Factors and Ergonomics Society* 48, no. 1 (2006): 196–205.

77   red light ahead   G. D. Logan, "An Instance Theory of Attention and Memory," *Psychological Review* 109 (2002): 376–400; D. L. Strayer and F. A. Drews,

"Attention," *Handbook of Applied Cognition,* ed. Francis T. Durso (Hoboken, N.J.: Wiley, 2007); A. D. Baddeley, "Selective Attention and Performance in Dangerous Environments," *British Journal of Psychology* 63 (1972): 537–46; E. Goldstein, *Cognitive Psychology: Connecting Mind, Research and Everyday Experience* (Independence, Ky.: Cengage Learning, 2014).

77 of common sense In response to a fact-checking email, Strayer expanded his comments: "With automated systems, we may not focus or concentrate attention on the task—we even mind wander in boring or repetitive settings. It takes effort to concentrate attention and this can lead to high levels of mental workload and we see a 'vigilance decrement' where attention lapses (and we make errors and miss critical events). This is often the case with monitoring tasks (keep an eye on the autonomous system) and when things go awry we may not notice or react on autopilot (even if this is not the correct action—we refer to this as slips where autopilot took over)."

78 gauges and controls Airbus, *Airbus A330 Aircraft Recovery Manual Airbus,* 2005, http://www.airbus.com/fileadmin/media_gallery/files/tech_data/ARM/ARM_A330_20091101.pdf.

80 throughout the flight The automatic warning system of this A330 was programmed so that the stall warning would cease when the plane's stall was most severe. In some situations, when the pitch attitude was too high and the airflow into the pitot tubes too low, the computer assumed the data it was gathering was erroneous. So it sounded no alarms. Thus, a perverse situation arose for Flight 447 after the pitot tubes thawed: At times, when Bonin did something to make the stall worse, the alarm stopped. The computers worked as programmed, but the result was information that might have been confusing to the pilots.

81 "reactive thinking" Koji Jimura, Maria S. Chushak, and Todd S. Braver, "Impulsivity and Self-Control During Intertemporal Decision Making Linked to the Neural Dynamics of Reward Value Representation," *The Journal of Neuroscience* 33, no. 1 (2013): 344–57; Ayeley P. Tchangani, "Modeling for Reactive Control and Decision Making in Uncertain Environment," in *Control and Learning in Robotic Systems,* ed. John X. Liu (New York: Nova Science Publishers, 2005), 21–58; Adam R. Aron, "From Reactive to Proactive and Selective Control: Developing a Richer Model for Stopping Inappropriate Responses," *Biological Psychiatry* 69, no. 12 (2011): 55–68; Veit Stuphorn and Erik Emeric, "Proactive and Reactive Control by the Medial Frontal Cortex," *Frontiers in Neuroengineering* 5 (2012): 9; Todd S. Braver et al., "Flexible Neural Mechanisms of Cognitive Control Within Human Prefrontal Cortex," *Proceedings of the National Academy of Sciences* 106, no. 18 (2009): 7351–56; Todd S. Braver,

"The Variable Nature of Cognitive Control: A Dual Mechanisms Framework." *Trends in Cognitive Sciences* 16, no. 2 (2012): 106–13; Yosuke Morishima, Jiro Okuda, and Katsuyuki Sakai, "Reactive Mechanism of Cognitive Control System," *Cerebral Cortex* 20, no. 11 (2010) 2675–83; Lin Zhiang and Kathleen Carley, "Proactive or Reactive: An Analysis of the Effect of Agent Style on Organizational Decision Making Performance," *Intelligent Systems in Accounting, Finance and Management* 2, no. 4 (1993): 271–87.

81 the psychologist, in 2009 Joel M. Cooper et al., "Shifting Eyes and Thinking Hard Keep Us in Our Lanes," *Human Factors and Ergonomics Society Annual Meeting Proceedings* 53, no. 23 (2009): 1753–56. For more on this topic, please see Frank A. Drews and David L. Strayer, "Chapter 11: Cellular Phones and Driver Distraction," in *Driver Distraction: Theory, Effects, and Mitigation*, ed. Michael A. Regan, John D. Lee, and Kristie L. Young (Boca Raton, Fla.: CRC Press, 2008): 169–90; Frank A. Drews, Monisha Pasupathi, and David L. Strayer, "Passenger and Cell Phone Conversations in Simulated Driving," *Journal of Experimental Psychology: Applied* 14, no. 4 (2008): 392; Joel M. Cooper, Nathan Medeiros-Ward, and David L. Strayer, "The Impact of Eye Movements and Cognitive Workload on Lateral Position Variability in Driving," *Human Factors: The Journal of the Human Factors and Ergonomics Society* 55, no. 5 (2013): 1001–14; David B. Kaber et al., "Driver Performance Effects of Simultaneous Visual and Cognitive Distraction and Adaptation Behavior," *Transportation Research Part F: Traffic Psychology and Behaviour* 15, no. 5 (2012): 491–501; I. J. Faulks et al., "Update on the Road Safety Benefits of Intelligent Vehicle Technologies—Research in 2008–2009," 2010 Australasian Road Safety Research, Policing and Education Conference, August 31–September 3, 2010, Canberra, Australia.

82 announcement of any kind In a fact-checking conversation, Stephen Casner, a research psychologist at NASA, said that if a plane was falling at ten thousand–plus feet per minute, the g-force would be pretty close to 1, and as a result, it would be unlikely the passengers would have noticed that anything was amiss. However, he added, "Actually, *no one* knows what that feels like. Everyone who has felt what it's like to lose 10,000 feet a minute dies pretty soon after feeling it."

83 ten thousand feet per minute In response to questions, a spokesman for Air France wrote: "A fundamental aspect is that the STALL alarm stopped when the speed fell below 60 kts, leading the pilots to think they were out of the stall. Especially that every time they pushed on the stick to try and get out of the stall situation, the STALL alarm started to work again, leading them to cancel their pitching action! Also, during the last phase, vertical speed indications were unstable, adding doubt and confusion in the pilots' minds."

85 Dayton, near where she lived In an email sent in reply to a fact-checking inquiry, Crandall wrote: "In 1986, I began working with Dr. Gary Klein at his company Klein Associates Inc. The work you mention with firefighters and military commanders had already begun when I joined the company. It continued for many years, expanding well beyond firefighting and military command and control, and was carried out by Gary and the Klein Associates research team (who were an amazing bunch of very smart talented quirky people). I had both research and management positions at Klein Associates, and I was involved in some of those studies, not in others. As owner and Chief Scientist, Gary led our efforts to describe how (some) people are able to 'keep their heads in chaotic environments' and particularly how (some) people are able to make effective decisions under conditions of stress, risk, and time pressure. . . . It is correct that in the interviews we conduct, when asked about decision making and how a person knew to do X in a particular situation, they often respond with, 'experience' or 'gut feel' or 'intuition' or 'I just knew.' . . . These accounts of an intuitive basis for decision making became a cornerstone of our research efforts. . . . The studies we did in the NICU confirmed what we were finding in other work domains—highly experienced, highly skilled personnel become very good at paying attention to what's most important (the critical cues) in a given situation, and not getting distracted by less important information. . . . Over time and repeated experience with similar situations, they learn what matters and what doesn't. They learn to size up a situation very quickly and accurately. They see connections across various cues (clusters; packages; linkages) that form a meaningful pattern. Some people refer to this as a gestalt, and others as 'mental models' or schemas." For more details, please see Beth Crandall and Karen Getchell-Reiter, "Critical Decision Method: A Technique for Eliciting Concrete Assessment Indicators from the Intuition of NICU Nurses," *Advances in Nursing Science* 16, no. 1 (1993): 42–51; B. Crandall and R. Calderwood, "Clinical Assessment Skills of Experienced Neonatal Intensive Care Nurses," *Contract* 1 (1989): R43; B. Crandall and V. Gamblian, "Guide to Early Sepsis Assessment in the NICU," *Instruction Manual Prepared for the Ohio Department of Development Under the Ohio SBIR Bridge Grant Program* (Fairborn, Ohio: Klein Associates, 1991).

87 "a whole picture" In an email sent in reply to a fact-checking inquiry, Crandall wrote: "The other nurse was a preceptee—in training to provide nursing care in a NICU. Darlene was her preceptor—helping her learn and providing oversight and guidance as she learns how to care for premature babies. So, the baby WAS Darlene's responsibility in the sense that she was supervising/precepting the nurse caring for the baby. You are correct, she noticed that the baby didn't look 'good.' Here is the incident account that we wrote up based on our interview notes: 'When this incident took place, I was teaching, serving as a preceptor for a new nurse. We had been working together for quite awhile and she was nearing the end of her orientation, so she was really doing primary care and I was in more of a supervisory position. Anyway, we were nearing the end of a shift and I walked by this particular isolette and the baby really caught my eye. The baby's color was off and its skin was mottled. Its belly looked slightly rounded. I looked at the chart and it indicated the baby's temp was unstable. I also noticed that the baby had had a heel stick for lab work several minutes ago and the stick was still bleeding. When I asked my orientee how she thought the baby was doing, she said that he seemed kind of sleepy to her. I went and got the Doctor immediately and told him we were "in big trouble" with this baby. I said the baby's temp was unstable, that its color was funny, it seemed lethargic and it was bleeding from a heel stick. He reacted right away, put the baby on antibiotics and ordered cultures done. I was upset with the orientee that she had missed these cues, or that she had noticed them but not put them together. When we talked about it later I asked about the baby's temp dropping over four readings. She had noticed it, but had responded by increasing the heat in the isolette. She had responded to the 'surface' problem, instead of trying to figure out what might be causing the problem."

88 **"creating mental models"** Thomas D. LaToza, Gina Venolia, and Robert DeLine, "Maintaining Mental Models: A Study of Developer Work Habits," *Proceedings of the 28th International Conference on Software Engineering* (New York: ACM, 2006); Philip Nicholas Johnson-Laird, "Mental Models and Cognitive Change," *Journal of Cognitive Psychology* 25, no. 2 (2013): 131–38; Philip Nicholas Johnson-Laird, *How We Reason* (Oxford: Oxford University Press, 2006); Philip Nicholas Johnson-Laird, *Mental Models,* Cognitive Science Series, no. 6 (Cambridge, Mass.: Harvard University Press, 1983); Earl K. Miller and Jonathan D. Cohen, "An Integrative Theory of Prefrontal Cortex Function," *Annual Review of Neuroscience* 24, no. 1 (2001): 167–202; J. D. Sterman and D. V. Ford, "Expert Knowledge Elicitation to Improve Mental and Formal Models," *Systems Approach to Learning and Education into the 21st Century,* vol. 1, 15th International System Dynamics Conference, August 19–22, 1997, Istanbul, Turkey; Pierre Barrouillet, Nelly Grosset, and Jean-François Lecas, "Conditional Reasoning by Mental Models: Chronometric and Developmental Evidence," *Cognition* 75, no. 3 (2000): 237–66; R. M. J. Byrne, *The Rational Imagination: How People Create Alternatives to Reality* (Cambridge, Mass.: MIT Press, 2005); P. C. Cheng and K. J. Holyoak, "Pragmatic Reasoning Schemas," in *Reasoning: Studies of Human Inference and Its Foundations,* eds. J. E. Adler and L. J. Rips (Cambridge: Cambridge University Press, 2008), 827–42; David P. O'Brien, "Human Reasoning Includes a Mental Logic," *Behavioral and Brain Sciences* 32, no. 1 (2009): 96–97; Niki Verschueren, Walter

Schaeken, and Gery d'Ydewalle, "Everyday Conditional Reasoning: A Working Memory–Dependent Tradeoff Between Counterexample and Likelihood Use," *Memory and Cognition* 33, no. 1 (2005): 107–19.

88 the child's bassinet In response to a fact-checking email, Crandall wrote: "The key to this story (for me anyway) is that experts see meaningful patterns that novices miss altogether. As an experienced NICU nurse, Darlene has seen hundreds of babies. She is not reflecting on all of them . . . they have merged into a sense of what is typical for a premie baby at X weeks. She has also seen many babies with sepsis (it happens a lot in NICUs, for a variety of reasons unrelated to quality of care). The combination of cues (bloody bandaid, falling temp, distended belly, sleepiness/lethargy) brought with it the recognition 'this baby is in trouble' and 'probably septic.' At least, that's what she told us in the interview. . . . I agree that people often create narratives to help explain what's going on around them, and help them make sense—particularly when they are having trouble figuring something out. In this incident, Darlene was not having trouble figuring out what was going on—she recognized immediately what was going on. . . . I think of Darlene's story as being about expertise, and the difference between how experts and novices view and understand a given situation. . . . Storytelling takes time, and stories are linear (this happened, then this, and then that). When experienced people describe events such as this one, what happens is very fast: They 'read' the situation, they understand what's going on, and they know what to do."

90 "It's even harder now" In response to a fact-checking email, Casner expanded his comments: "I wouldn't say that pilots are 'passive' but that they find it exceedingly difficult to maintain their attention on an automated system that works so reliably well. Humans are not good at sitting and staring. . . . Humans have limited attentional resources (e.g., how our kids do stuff behind our backs and get away with it). So we have to keep our attention pointed in the direction that we think is most important at all times. If a cockpit computer in front of me has worked impeccably for 100 hours in a row, it's hard to envision that as being the most important thing to think about. For example, my kid could be getting away with some insane stuff at that very moment. In our study of mind wandering among pilots [*Thoughts in Flight: Automation Use and Pilots' Task-Related and Task-Unrelated Thought*], we found that the pilot flying was thinking 'task-unrelated thoughts' about 30% of the time. The other pilot, the monitoring pilot, was mind wandering about 50% of the time. Why wouldn't they? If you don't give me something important or pressing to think about, I'll come up with something myself."

90 people build mental models Sinan Aral, Erik Brynjolfsson, and Marshall Van Alstyne, "Information, Technology, and Information Worker Productivity," *Information Systems Research* 23, no. 3 (2012): 849–67; Sinan Aral and

Marshall Van Alstyne, "The Diversity-Bandwidth Trade-Off," *American Journal of Sociology* 117, no. 1 (2011): 90–171; Nathaniel Bulkley and Marshall W. Van Alstyne, "Why Information Should Influence Productivity" (2004); Nathaniel Bulkley and Marshall W. Van Alstyne, "An Empirical Analysis of Strategies and Efficiencies in Social Networks," Boston U. School of Management research paper no. 2010-29, MIT Sloan research paper no. 4682-08, February 1, 2006, http://ssrn.com/abstract=887406; Neil Gandal, Charles King, and Marshall Van Alstyne, "The Social Network Within a Management Recruiting Firm: Network Structure and Output," *Review of Network Economics* 8, no. 4 (2009): 302–24.

90 leveraged existing skills In response to a fact-checking email, Van Alstyne expanded upon his comments: "One of the original hypotheses attributed the gains of the smaller project load to the efficacy associated with economies of specialization. Doing a singular, focused activity can make you very good at that activity. The idea goes all the way back to Adam Smith and the efficiency associated with focused tasks at a pin factory. Generalization, or pursuing diverse work in our context, meant spreading projects across finance, education, and commercial IT. These are very different industries. Running projects across them requires different knowledge and it also means tapping different social networks. Specialization, in these consulting projects, meant focusing on, say, just the finance projects. Knowledge could be deepened within this focal area and the social network could be adapted to finance contacts alone. At least this is one theory as to why specialization might be better. Obviously, specialization can restrict the number of possible projects—there might not be a new finance project when there does happen to be one, or several, in education or IT. But perhaps if you wait, you'll get another finance project."

91 deemed a success In response to a fact-checking email, Van Alstyne identified other reasons why joining small numbers of projects, and a project at its start, had benefits: "The first is multitasking. Initially, taking on new projects strictly increases output, in this case revenues generated by these consultants. Revenue growth can continue even past the point where the productivity on a given project starts to fall. Consider a project as a collection of tasks (assessing client needs, generating target candidates, selecting candidates, vetting resumes, presenting options to clients, closing the deal . . .). As a person takes on new work, its tasks displace some tasks of the existing work. So an existing project can take longer when a person takes on a new project, drawing out the period over which he/she gets paid. Total throughput, however, can still rise for awhile as a person takes on new projects. The stream of revenues brought in by a person juggling 6 projects tends to be higher than the stream of revenues brought in by a person juggling 4 even though each of the 6 projects takes longer than it would have taken if it were only in a group of 4. At some point, however, this relationship trends completely downward. New projects take too long *and* revenues decline. Taking on another project strictly decreases productivity. As one consultant put it, 'There are too many balls in the air and then too many get dropped.' It takes too long to complete tasks, some tasks are not completed at all, and the flow of revenues dribbles out over a really long period. So there is an optimal number of projects to take on and this is below 12. The second consideration, as you suggest, is access to rich information. This exhibits a similar invert-U pattern. We were able to judge how much novel information each person received by tracking their actual email communication. We measured this both in a sense of 'variance,' i.e., how *unusual* was a fact relative to other received facts, and also in terms of 'volume,' i.e., how *many* new facts a person received. . . . Initially, greater access to more novel information strictly increased productivity. Superstars did receive about 25% more novel information than their typical peer and this access to novelty helped predict their success. Eventually, however, those outlying people who received the absolute highest novelty—about twice that of the superstars—were less productive than the superstars. Either excess information was too weird, off-topic, and not actionable or excess information was too much to process. A massive volume of novelty introduces the white-collar worker's equivalent of the 'Where's Waldo' problem: You can't find the important information in all the noise. Both of these factors were statistically significant predictors of the superstars."

93   bright morning sky Richard De Crespigny, *QF32* (Sydney: Pan Macmillan Australia, 2012); *Aviation Safety Investigation Report 089: In-Flight Uncontained Engine Failure Airbus A380–842, VH-OQA* (Canberra: Australian Transport Safety Bureau, Department of Transport and Regional Services, 2013); Jordan Chong, "Repaired Qantas A380 Arrives in Sydney," *The Sydney Morning Herald,* April 22, 2012; Tim Robinson, "Qantas QF32 Flight from the Cockpit," *The Royal Aeronautical Society,* December 8, 2010; "Qantas Airbus A380 Inflight Engine Failure," Australian Transport Safety Bureau, December 8, 2010; "Aviation Occurrence Investigation AO-2010–089 Interim-Factual," Australian Transport Safety Bureau, May 18, 2011; "In-Flight Uncontained Engine Failure—Overhead Batam Island, Indonesia, November 4, 2010, VH-OQA, Airbus A380–842," Australian Transport Safety Bureau, investigation no. AO-2010–089, Sydney.

95   de Crespigny later told me I am indebted to Captain de Crespigny for his time as well as his book, *QF32.* In an interview, de Crespigny emphasized that he is speaking for himself, and not for Qantas, in recalling and describing these events.

99 **"models they can use"** In response to a fact-checking email, Burian expanded upon her comments and said that her comments should be read in the light of "shifting focus from what was wrong/malfunctioning/not available to what was working/functioning/available was a turning point. I spoke of how this happened for him in this specific situation but generalized to how this shift in mindset has been found to be quite helpful to pilots, particularly when faced with multiple failure conditions. . . . Modern aircraft are highly technically advanced and their system designs are tightly coupled and fairly opaque. This can make it quite difficult for pilots to understand the whys and wherefores of some malfunctions and how multiple malfunctions might be associated with each other. Instead of trying to sort through a myriad of malfunctions and think about how they are connected and the implications they have, shifting focus to an aircraft's capabilities simplifies the cognitive demands and can facilitate deciding how to do what is needing to be done. . . . Once a critical event has occurred, really good pilots do several things—they try to determine what is most critical to be dealt with first (narrowing of attention) but also pull back from time to time (broadening of attention) to do two things: 1) make sure they are not missing cues/information that might contradict or alter their understanding of their situation and 2) track the overall situation as part of their assessment of the most critical things to be attending to. For example, consider a catastrophic emergency (requiring an emergency landing/ditching) that occurs at cruise altitude. The crew will have some time to deal with the condition, but at some point, their attention should shift from dealing directly with the malfunction/condition to preparing for and executing a ditching/landing. Good pilots are constantly assessing the actions being taken, their efficacy, and needed actions relative to the overall status of the aircraft and phase of flight. Of course, good pilots also fully enlist the help of others in doing all this (i.e., good CRM). Good pilots also do a lot of 'what if' exercises before any event occurs, mentally running through a variety of scenarios to think about what they might do, how the situation might unfold, circumstances that would alter the way(s) in which they would respond, etc. General aviation pilots are taught to do something similar during flight when they say to themselves at various points along their route 'If I were to lose my (only) engine right now (i.e., engine dies), where would I land?' "

99 **"land the plane"** In response to a fact-checking email, de Crespigny expanded upon his comments: "Dave used [an onboard computer] program to check the landing distance. His first pass resulted in NO SOLUTION because there were too many failures for the program to come up with a landing solution. Dave then simplified the entries for the failures. The LDPA program [the landing distance performance application] then displayed a landing distance margin of just 100 metres. Whilst Dave and the others were calculating the performance (that turned out to be incorrect anyways because of errors in the LDPA program and more extensive aircraft (brakes) damage than what was reported), I kept a broad situation awareness of the entire operation: aircraft, fuel, critical paths, pilot duties, cabin crew, passengers, air traffic control, emergency services. . . . Simplifying the A380 (with 4,000 parts) down to a Cessna (the flying version of the 1938 Ariel Red Hunter motorcycle) kept things very simple for me, removing the complexity, making each system simple to understand from a mechanical (not mechatronic perspective), simplifying my mental model of the aircraft's systems, freeing up mind-space to manage the entire event. It [is] vital in an emergency that there is a structured hierarchy of responsibility and authority. It's even more important that pilots understand the roles, tasks, and teamwork required in an autonomous team of just two pilots (more in our case on board QF32), isolated from help but in charge of 469 lives."

101　fail every time In response to a fact-checking email, de Crespigny explained that it is impossible to get a simulator to re-create the conditions of QF32, because the problems with the plane were so extreme.

CHAPTER FOUR: GOAL SETTING

103　about to attack For my understanding of the events leading up to the Yom Kippur War, I am indebted to Professor Uri Bar-Joseph, who was kind enough to provide extensive written comments, as well as the following sources: Abraham Rabinovich, *The Yom Kippur War: The Epic Encounter That Transformed the Middle East* (New York: Schocken, 2007); Uri Bar-Joseph, *The Watchman Fell Asleep: The Surprise of Yom Kippur and Its Sources* (Albany: State University of New York Press, 2012); Uri Bar-Joseph, "Israel's 1973 Intelligence Failure," *Israel Affairs* 6, no. 1 (1999): 11–35; Uri Bar-Joseph and Arie W. Kruglanski, "Intelligence Failure and Need for Cognitive Closure: On the Psychology of the Yom Kippur Surprise," *Political Psychology* 24, no. 1 (2003): 75–99; Yosef Kuperwaser, *Lessons from Israel's Intelligence Reforms* (Washington, D.C.: Saban Center for Middle East Policy at the Brookings Institution, 2007); Uri Bar-Joseph and Jack S. Levy, "Conscious Action and Intelligence Failure," *Political Science Quarterly* 124, no. 3 (2009): 461–88; Uri Bar-Joseph and Rose McDermott, "Personal Functioning Under Stress Accountability and Social Support of Israeli Leaders in the Yom Kippur War," *Journal of Conflict Resolution* 52, no. 1 (2008): 144–70; Uri Bar-Joseph, " 'The Special Means of Collection': The Missing Link in the Surprise of the Yom Kippur War," *The Middle East Journal* 67, no. 4 (2013): 531–46; Yaakov Lapin, "Declassified Yom Kippur War Papers Reveal Failures," *The Jerusalem Post*, September 20,

2012; Hamid Hussain, "Opinion: The Fourth Round—A Critical Review of 1973 Arab-Israeli War," *Defence Journal*, November 2002, http://www.defence journal.com/2002/nov/4th-round.htm; P. R. Kumaraswamy, *Revisiting the Yom Kippur War* (London: Frank Cass, 2000); Charles Liebman, "The Myth of Defeat: The Memory of the Yom Kippur War in Israeli Society," *Middle Eastern Studies* 29, no. 3 (1993): 411; Simon Dunstan, *The Yom Kippur War: The Arab-Israeli War of 1973* (Oxford: Osprey Publishing, 2007); Asaf Siniver, *The Yom Kippur War: Politics, Legacy, Diplomacy* (Oxford: Oxford University Press, 2013).

104 **"sharp as possible"** Bar-Joseph, *Watchman Fell Asleep.*

105 **nothing more than words** In an email, the historian Uri Bar-Joseph wrote that the concept was "a set of assumptions that were based on documented information that was passed to Israel by Ashraf Marwan, the son-in-law of late president Nasser and a close advisor to Sadat, who since late 1970 worked for the Mossad. The main assumptions were: (1) Egypt cannot occupy the Sinai without neutralizing the Israeli air-superiority. The way to do it is by attacking the bases of the [Israeli Air Force] at the beginning of the war. In order to do it, Egypt needs long-range attack aircraft which she won't have before 1975; (2) In order to deter Israel from attacking strategic targets in Egypt, Egypt needs Scud missiles that will be able to hit Tel Aviv. Scuds started arriving in Egypt in the summer of 1973 but were not expected to be operational before February 1974. (3) Syria will not go to war without Egypt. Zeira became an ardent believer in these assumptions and turned them into an orthodox conception, which he kept until war started."

106 **within the next decade** Bar-Joseph and Kruglanski, "Intelligence Failure and Need for Cognitive Closure," 75–99.

107 **need for cognitive closure** For more on cognitive closure, please see Steven L. Neuberg and Jason T. Newsom, "Personal Need for Structure: Individual Differences in the Desire for Simpler Structure," *Journal of Personality and Social Psychology* 65, no. 1 (1993): 113; Cynthia T. F. Klein and Donna M. Webster, "Individual Differences in Argument Scrutiny as Motivated by Need for Cognitive Closure," *Basic and Applied Social Psychology* 22, no. 2 (2000): 119–29; Carsten K. W. De Dreu, Sander L. Koole, and Frans L. Oldersma, "On the Seizing and Freezing of Negotiator Inferences: Need for Cognitive Closure Moderates the Use of Heuristics in Negotiation," *Personality and Social Psychology Bulletin* 25, no. 3 (1999): 348–62; A. Chirumbolo, A. Areni, and G. Sensales, "Need for Cognitive Closure and Politics: Voting, Political Attitudes and Attributional Style," *International Journal of Psychology* 39 (2004): 245–53; Arie W. Kruglanski, *The Psychology of Closed Mindedness* (New York: Psychology Press, 2013); Arie W. Kruglanski et al., "When Similarity Breeds Content: Need for Closure and the Allure of Homogeneous and Self-Resembling Groups," *Journal of Personality and Social Psychology* 83, no. 3 (2002): 648; Steven L. Neuberg and Jason T. Newsom, "Personal Need for Structure: Individual Differences in the Desire for Simpler Structure," *Journal of Personality and Social Psychology* 65, no. 1 (1993): 113.

107 "confusion and ambiguity" Bar-Joseph, *Watchman Fell Asleep;* Donna M. Webster and Arie W. Kruglanski, "Individual Differences in Need for Cognitive Closure," *Journal of Personality and Social Psychology* 67, no. 6 (1994): 1049.

108 "need for closure introduces a bias" Bar-Joseph and Kruglanski, "Intelligence Failure and Need for Cognitive Closure," 75–99.

108 Donna Webster, wrote in 1996 Arie W. Kruglanski and Donna M. Webster, "Motivated Closing of the Mind: 'Seizing' and 'Freezing,'" *Psychological Review* 103, no. 2 (1996): 263.

109 it has been selected Ibid.; De Dreu, Koole, and Oldersma, "On the Seizing and Freezing of Negotiator Inferences," 348–62.

109 we're making a mistake In an email responding to fact-checking questions, Arie Kruglanski wrote: "People under high need for closure have trouble appreciating others' perspectives and points of view. People under high need for closure also prefer hierarchical, autocratic, decision making structures in groups because those provide better closure than horizontal or democratic structures that tend to be more chaotic. People under high need for closure are therefore intolerant of diversity, and of dissent in groups and aren't very creative. Politically, conservatives tend to be higher on need for closure than liberals, but people with high need for closure tend to be more committed to things and values than people low on need for closure."

110 "should not expect promotion" Bar-Joseph and Kruglanski, "Intelligence Failure and Need for Cognitive Closure," 75–99.

110 "outside the organization" Uri Bar-Joseph, "Intelligence Failure and Success in the War of Yom Kippur," unpublished paper.

113 "before war broke out" Abraham Rabinovich, "Three Years Too Late, Golda Meir Understood How War Could Have Been Avoided," *The Times of Israel,* September 12, 2013.

115 Israelis were killed or wounded Zeev Schiff, *A History of the Israeli Army, 1874 to the Present* (New York: Macmillan, 1985).

115 "generation was nearly lost" Richard S. Lazarus, *Fifty Years of the Research and Theory of RS Lazarus: An Analysis of Historical and Perennial Issues* (New York: Psychology Press, 2013).

115 **"Even a quarter century later"** Kumaraswamy, *Revisiting the Yom Kippur War.*

115 **good at choosing goals** For my understanding of General Electric, I am indebted to Joseph L. Bower and Jay Dial, "Jack Welch: General Electric's Revolutionary," Harvard Business School case study no. 394-065, October 1993, revised April 1994; Francis Aguilar and Thomas W. Malnight, "General Electric Co: Preparing for the 1990s," Harvard Business School case study no. 9-390, December 20, 1989; Francis J. Aguilar, R. Hamermesh, and Caroline Brainard, "General Electric: Reg Jones and Jack Welch," Harvard Business School case study no. 9-391-144, June 29, 1991; Kirsten Lungberg, "General Electric and the National Broadcasting Company: A Clash of Cultures," Harvard University John F. Kennedy School of Government case study, 1989; Nitin Nohria, Anthony J. Mayo, and Mark Benson, "General Electric's 20th Century CEOs," Harvard Business School case study, December 2005; Jack Welch and John A. Byrne, *Jack: Straight from the Gut* (New York: Warner, 2003); Larry Greiner, "Steve Kerr and His Years with Jack Welch at GE," *Journal of Management Inquiry* 11, no. 4 (2002): 343–50; Stratford Sherman, "The Mind of Jack Welch," *Fortune,* March 27, 1989; Marilyn Harris et al., "Can Jack Welch Reinvent GE?" *BusinessWeek,* June 30, 1986; Mark Potts, "GE Chief Hopes to Shape Agile Giant," *Los Angeles Times,* June 1, 1988; Noel Tichy and Ram Charan, "Speed Simplicity and Self-Confidence: An Interview with Jack Welch," *Harvard Business Review,* September 1989; Ronald Grover and Mark Landler, "NBC Is No Longer a Feather in GE's Cap," *BusinessWeek,* June 2, 1991; Harry Bernstein, "The Two Faces of GE's 'Welchism,'" *Los Angeles Times,* January 12, 1988; "Jack Welch Reinvents General Electric. Again," *The Economist,* March 30, 1991; L. J. Dans, "They Call Him 'Neutron,'" *Business Month,* March 1988; Richard Ellsworth and Michael Kraft, "Jack Welch at GE: 1981–1989," Claremont Graduate School, Peter F. Drucker and Masatoshi Ito Graduate School of Management case study; Peter Petre, "Jack Welch: The Man Who Brought GE to Life," *Fortune,* January 5, 1987; Peter Petre, "What Welch Has Wrought at GE," *Fortune,* July 7, 1986; Stephen W. Quickel, "Welch on Welch," *Financial World,* April 3, 1990; Monica Roman, "Big Changes Are Galvanizing General Electric," *BusinessWeek,* December 18, 1989; Thomas Stewart, "GE Keeps Those Ideas Coming," *Fortune,* August 12, 1991.

116 **"became the work 'contract'"** Nitin Nohria, Anthony J. Mayo, and Mark Benson, "General Electric's 20th Century CEOs," *Harvard Business Review,* December 19, 2005, revised April 2011; John Cunningham Wood and Michael C. Wood, *Peter F. Drucker: Critical Evaluations in Business and Management,* vol. 1 (London: Routledge, 2005).

117 **best way to set goals** Gary P. Latham, Terence R. Mitchell, and Dennis L. Dossett, "Importance of Participative Goal Setting and Anticipated Rewards on Goal Difficulty and Job Performance," *Journal of Applied Psychology* 63, no. 2 (1978): 163; Gary P. Latham and Gerard H. Seijts, "The Effects of Proximal and Distal Goals on Performance on a Moderately Complex Task," *Journal of Organizational Behavior* 20, no. 4 (1999): 421–29; Gary P. Latham and J. James Baldes, "The 'Practical Significance' of Locke's Theory of Goal Setting," *Journal of Applied Psychology* 60, no. 1 (1975): 122; Gary P. Latham and Craig C. Pinder, "Work Motivation Theory and Research at the Dawn of the Twenty-First Century," *Annual Review of Psychology* 56 (2005): 485–516; Edwin A. Locke and Gary P. Latham, "Building a Practically Useful Theory of Goal Setting and Task Motivation: A Thirty-Five-Year Odyssey," *American Psychologist* 57, no. 9 (2002): 705; A. Bandura, "Self-Regulation of Motivation and Action Through Internal Standards and Goal Systems," in *Goal Concepts in Personality and Social Psychology*, ed. L. A. Pervin (Hillsdale, N.J.: Erlbaum, 1989), 19–85; Travor C. Brown and Gary P. Latham, "The Effects of Goal Setting and Self-Instruction Training on the Performance of Unionized Employees," *Relations Industrielles/Industrial Relations* 55, no. 1 (2000): 80–95; Judith F. Bryan and Edwin A. Locke, "Goal Setting as a Means of Increasing Motivation," *Journal of Applied Psychology* 51, no. 3 (1967): 274; Scott B. Button, John E. Mathieu, and Dennis M. Zajac, "Goal Orientation in Organizational Research: A Conceptual and Empirical Foundation," *Organizational Behavior and Human Decision Processes* 67, no. 1 (1996): 26–48; Dennis L. Dossett, Gary P. Latham, and Terence R. Mitchell, "Effects of Assigned Versus Participatively Set Goals, Knowledge of Results, and Individual Differences on Employee Behavior When Goal Difficulty Is Held Constant," *Journal of Applied Psychology* 64, no. 3 (1979): 291; Elaine S. Elliott and Carol S. Dweck, "Goals: An Approach to Motivation and Achievement," *Journal of Personality and Social Psychology* 54, no. 1 (1988): 5; Judith M. Harackiewicz et al., "Predictors and Consequences of Achievement Goals in the College Classroom: Maintaining Interest and Making the Grade," *Journal of Personality and Social Psychology* 73, no. 6 (1997): 1284; Howard J. Klein et al., "Goal Commitment and the Goal-Setting Process: Conceptual Clarification and Empirical Synthesis," *Journal of Applied Psychology* 84, no. 6 (1999): 885; Gary P. Latham and Herbert A. Marshall, "The Effects of Self-Set, Participatively Set, and Assigned Goals on the Performance of Government Employees," *Personnel Psychology* 35, no. 2 (June 1982): 399–404; Gary P. Latham, Terence R. Mitchell, and Dennis L. Dossett, "Importance of Participative Goal Setting and Anticipated Rewards on Goal Difficulty and Job Performance," *Journal of Applied Psychology* 63, no. 2 (1978): 163; Gary P. Latham and Lise M. Saari, "The Effects of Holding Goal Difficulty Constant on Assigned and Participatively Set Goals," *Academy of Management Journal* 22, no. 1 (1979): 163–68; Don VandeWalle, William L. Cron, and John W. Slocum, Jr., "The Role of Goal Orientation Following Performance Feedback," *Journal of Applied Psychology* 86, no. 4 (2001): 629; Edwin A. Locke and Gary P. Latham, eds., *New Developments in Goal Setting and Task Performance* (London: Routledge, 2013).

117 how fast they produced text Gary P. Latham and Gary A. Yukl, "Assigned Versus Participative Goal Setting with Educated and Uneducated Woods Workers," *Journal of Applied Psychology* 60, no. 3 (1975): 299.

118 "Making yourself break a goal" In an email responding to fact-checking questions, Latham wrote that achieving goals also requires access to the necessary resources and feedback on goal progress. "For long-term/distal goals, proximal/sub goals should be set. Sub goals do two things: maintain motivation for attaining the distal goal as the attainment of one sub goal leads to the desire to attain another sub goal. Second, feedback from pursuit of each sub goal yields information as to whether you are on- or off-track."

120 Latham wrote in 1990 Edwin A. Locke and Gary P. Latham, "New Directions in Goal-Setting Theory," *Current Directions in Psychological Science* 15, no. 5 (2006): 265–68.

120 "the right things," said Latham In an email responding to fact-checking questions, Latham wrote: "When people lack the ability to attain a performance goal, that is, a goal having to do with a specific desired result such as a golf score of 80 or a 23% increase in revenue, [improper focus or tunnel vision] may occur. The solution is to set a specific, challenging learning goal where the emphasis is on discovering/developing a process, procedure, system that will enable you to improve your performance such as [coming] up with 5 ways you can improve your putting as opposed to put the ball in the cup in no more than 2 strokes."

121 business school, for help Kerr was initially one of twenty-four consultants brought in by Jack Welch to expand Work-Outs throughout GE.

121 more long-term plans Noel M. Tichy and Stratford Sherman, "Walking the Talk at GE," *Training and Development* 47, no. 6 (1993): 26–35; Ronald Henkoff, "New Management Secrets from Japan," *Fortune*, November 27, 1995; Ron Ashkenas, "Why Work-Out Works: Lessons from GE's Transformation Process," *Handbook of Business Strategy* 4, no. 1 (2003): 15–21; Charles Fishman, "Engines of Democracy," *Fast Company*, October 1999, http://www .fastcompany.com/37815/engines-democracy; Thomas A. Stewart, "GE Keeps Those Ideas Coming," in Rosabeth Moss Kanter, Barry A. Stein, and Todd D. Jick, *The Challenge of Organizational Change: How Companies Experience It and Leaders Guide It* (New York: The Free Press, 1992): 474–482; Joseph P. Cosco, "General Electric Works It All Out," *Journal of Business Strategy* 15, no. 3 (1994): 48–50.

121 **"turn out great"** In an email responding to fact-checking questions, Kerr wrote: "I stressed to the leadership teams that 'saying no to a bad idea is as useful as saying yes to a good one,' but that they couldn't dismiss any recommendation by saying things like: 'We thought of that already,' or 'We tried it before and it didn't work.' I always made the point that Work-Outs present a terrific opportunity to teach people about the business, and that they owed people a professional, courteous explanation as to why they didn't support a particular recommendation."

121 **SMART criteria** In an email responding to fact-checking questions, Kerr wrote that he never encouraged people to submit proposals without a rough plan and timeline. "The details of the plan would have to be sketched out after approval," he wrote.

122 **"ideas are fair game"** Cosco, "General Electric Works It All Out," 48–50.

123 **Japan's railway system** Ronald Henkoff, "New Management Secrets from Japan," *Fortune*, November 27, 1995.

123 **invent a faster train** The story of Japan's bullet train as it was told to Jack Welch (and has been repeated in popular nonfiction) differs slightly from the historical record. The account given here reflects the story that was told to Welch, but there are some details that story did not include, such as the fact that the concept for high-speed rail was explored but then abandoned by the Japanese railway prior to World War II. In an email responding to fact-checking questions, a representative of the Central Japan Railway Company wrote that in the 1950s the "Tokaido Line, the main line of Japan, was very crowded and [passengers had] been increasing because of the economical growth after the war, and Japan had to meet the growing needs of passengers to move between Tokyo (capital and largest city) and Osaka (second largest city). Actually there was a concept of 'Bullet train' before the WWII, [in] 1939 . . . but because of the war, that plan [had] been suspended. Japan National Railway decided to build [a] new line by standard gauge (many of Japanese conventional [lines adopted] narrow gauge) in 1957. The plan [was accepted] in 1958 by the government and construction had started." It is also worth noting that private efforts at developing faster trains were also occurring at the same time in Japan. The Odakyu Electric Railway, for instance, was developing a train capable of going ninety miles per hour. For a better understanding of the history of the bullet train, I recommend Toshiji Takatsu, "The History and Future of High-Speed Railways in Japan," *Japan Railway and Transport Review* 48 (2007): 6–21; Mamoru Taniguchi, "High Speed Rail in Japan: A Review and Evaluation of the Shinkansen Train" (working paper no. UCTC 103, University of California Transportation Center, 1992); Roderick Smith, "The Japanese Shinkansen: Catalyst for the

Renaissance of Rail," *The Journal of Transport History* 24, no. 2 (2003): 222–37; Moshe Givoni, "Development and Impact of the Modern High-Speed Train: A Review," *Transport Reviews* 26, no. 5 (2006): 593–611.

123   120 miles per hour In an email responding to fact-checking questions, a representative of the Central Japan Railway Company wrote that "in Japan, [a] JNR (Japan National Railway) engineer was considered [the] elite of Japanese engineers at that time, and the engineer who designed Shinkansen (Mr. Shima) was one of the engineers of JNR. . . . He [had] been working in JNR [a] long time already and had knowledge and experience about railways." Mr. Shima, the spokesperson noted, was asked, starting in 1955, to oversee Tōkaidō Shinkansen. "At the time of the bullet train project in 1939 I mentioned before, they were already planning to design trains which have [a max speed of] 125 mph. [The] engineer of Shinkansen had the clear aim of tying Tokyo to Osaka by 3 hours from the beginning, and [the] prototype called 'Series 1000' achieved 256 km/h (160 mph) in 1963."

124   into the 1980s Andrew B. Bernard, Andreas Moxnes, and Yukiko U. Saito, *Geography and Firm Performance in the Japanese Production Network* (working paper no. 14034, National Bureau of Economic Research, 2014).

125   "bullet train thinking" S. Kerr and S. Sherman, "Stretch Goals: The Dark Side of Asking for Miracles," *Fortune,* November 13, 1995; Sim B. Sitkin et al., "The Paradox of Stretch Goals: Organizations in Pursuit of the Seemingly Impossible," *Academy of Management Review* 36, no. 3 (2011): 544–66; Scott Jeffrey, Alan Webb, and Axel K-D. Schulz, "The Effectiveness of Tiered Goals Versus Stretch Goals," CAAA 2006 Annual Conference Paper (2006); Kenneth R. Thompson, Wayne A. Hochwarter, and Nicholas J. Mathys, "Stretch Targets: What Makes Them Effective?" *The Academy of Management Executive* 11, no. 3 (1997): 48–60; S. Kerr and D. LePelley, "Stretch Goals: Risks, Possibilities, and Best Practices," *New Developments in Goal Setting and Task Performance* (2013): 21–31; Steven Kerr and Steffen Landauer, "Using Stretch Goals to Promote Organizational Effectiveness and Personal Growth: General Electric and Goldman Sachs," *The Academy of Management Executive* 18, no. 4 (2004): 134–38; Kelly E. See, "Motivating Individual Performance with Challenging Goals: Is It Better to Stretch a Little or a Lot?" (manuscript presented for publication, Duke University, June 2003); Adrian D. Manning, David B. Lindenmayer, and Joern Fischer, "Stretch Goals and Backcasting: Approaches for Overcoming Barriers to Large-Scale Ecological Restoration," *Restoration Ecology* 14, no. 4 (2006): 487–92; Jim Heskett, "Has the Time Come for 'Stretch' in Management?" Harvard Business School, *Working Knowledge,* August 1, 2008, http://hbswk.hbs.edu/item/5989.html.

126 **their own workflow** Fishman, "Engines of Democracy," 33.

126 **goal would have done that** In an email responding to fact-checking questions, a spokesman for General Electric wrote that "the Durham plant was created with the flexibility to make such dramatic change[s]. Many adjustments were in process when the plant was opened in 1992. Durham from its inception was created as an 'incubator' for new manufacturing practices at GE Aviation. Yes, Jack [Welch] set the bar high—but given the aggressive competition in the aviation business, these goals were a requirement to be successful and to generate the kind of income necessary to fund new engine developments at that time (namely the GE90)."

127 **throughout the firm** Thompson, Hochwarter, and Mathys, "Stretch Targets," 48–60.

127 **Thinsulate** William E. Coyne, "How 3M Innovates for Long-Term Growth," *Research-Technology Management* 44, no. 2 (2001): 21–24.

127 **"broad search, or playfulness"** Sitkin et al., "Paradox of Stretch Goals," 544–66.

128 **the researchers wrote** Jeffrey, Webb, and Schulz, "The Effectiveness of Tiered Goals Versus Stretch Goals."

128 **University of Waterloo** Ibid.

128 **University of Melbourne** Thompson, Hochwarter, and Mathys, "Stretch Targets," 48–60.

129 **our in-box** Gil Yolanda et al., "Capturing Common Knowledge About Tasks: Intelligent Assistance for To-Do Lists," *ACM Transactions on Interactive Intelligent Systems (TiiS)* 2, no. 3 (2012): 15; Victoria Bellotti et al., "What a To-Do: Studies of Task Management Towards the Design of a Personal Task List Manager," *Proceedings of the SIGCHI Conference on Human Factors in Computing Systems* (2004): 735–42; Gabriele Oettingen and Doris Mayer, "The Motivating Function of Thinking About the Future: Expectations Versus Fantasies," *Journal of Personality and Social Psychology* 83, no. 5 (2002): 1198; Anja Achtziger et al., "Metacognitive Processes in the Self-Regulation of Goal Pursuit," in *Social Metacognition,* ed. Pablo Briñol and Kenneth DeMarree, Frontier of Social Psychology series (New York: Psychology Press, 2012), 121–39.

131 **throughout corporate America** Critics of stretch goals say that, if unconstrained, they can negatively impact an organization. For more, please see Lisa D. Ordóñez et al., "Goals Gone Wild: The Systematic Side Effects of Overprescribing Goal Setting," *The Academy of Management Perspectives* 23, no. 1 (2009): 6–16. And the response of Edwin A. Locke and Gary P. Latham, "Has Goal Setting Gone Wild, or Have Its Attackers Abandoned Good Scholarship?" *The Academy of Management Perspectives* 23, no. 1 (2009): 17–23.

131 **investigators concluded** The Commission of Inquiry, *The Yom Kippur War, an Additional Partial Report: Reasoning and Complement to the Partial Report of April 1, 1974*, vol. 1 (Jerusalem: 1974).

133 **all to blame** Mitch Ginsberg, "40 Years On, Yom Kippur War Intel Chiefs Trade Barbs," *The Times of Israel*, October 6, 2013; "Eli Zeira's Mea Culpa," *Haaretz*, September 22, 2004; Lilach Shoval, "Yom Kippur War Intelligence Chief Comes Under Attack 40 Years Later," *Israel Hayom*, October 7, 2013.

133 **"You are lying!"** Ibid.

CHAPTER FIVE: MANAGING OTHERS

134 **shoo them away** As mentioned in the chapter, the Federal Bureau of Investigation and Frank, Christie, and Colleen Janssen were all provided with summaries of this chapter and asked to respond to the details of this reporting. The FBI declined to comment, except as specified below. The Janssen family did not reply to repeated attempts to seek their comments by telephone and mail. The sources used in reporting details of the Janssen case include interviews as well as documents from *United States of America v. Kelvin Melton, Quantavious Thompson, Jakym Camel Tibbs, Tianna Daney Maynard, Jenna Martin, Clifton James Roberts, Patricia Ann Kramer, Jevante Price, and Michael Martell Gooden* (nos. 5:14-CR-72–1; 5:14-CR-72–2; 5:14-CR-72–3; 5:14-CR-72–4; 5:14-CR-72–5; 5:14-CR-72–6; 5:14-CR-72–7; 5:14-CR-72–8; 5:14-CR-72–9), filed in the U.S. District Court for the Eastern District of North Carolina Western Division; Affidavit in Support of Application for a Court Order Approving Emergency Interceptions, in the Matter of the Application of the United States of America for an Order Authorizing the Interception of Wire and Electronic Communications, no. 5:14-MJ-1315-D, filed in the U.S. District Court Eastern District of North Carolina Western Division; *United States v. Kelvin Melton*, Criminal Case no. 5:14-MJ-1316, filed in the U.S. District Court Eastern District of North Carolina; *United States v. Clifton James Roberts*, Criminal Case no. 5:14-MJ-1313, filed in the U.S. District Court Eastern District of North Carolina; *United States v. Chason Renee Chase, a/k/a "Lady Jamaica,"* Criminal Case no. 3:14-MJ-50, filed in the U.S. District Court for the District of South Carolina, and other court filings related to the alleged Janssen abduction. Details also came from Alan G. Breed and Michael Biesecher, "FBI: NC Inmate Helped Orchestrate Kidnapping," Associated Press, April 11, 2014; Kelly Gardner, "FBI Now Investigating Wake Forest Man's Disappearance," WRAL.com, April 8, 2014; Alyssa Newcomb, "FBI Rescued Kidnap Victim as Suspects Discussed Killing Him, Feds Say," *Good Morning America*, April 10, 2014; Anne Blythe and Ron Gallagher, "FBI Rescues Wake Forest Man; Abduction Related to Daughter's Work as Prosecutor, Investigators Say," *The Charlotte Observer*, April 10, 2014; Michael Biesecher and Kate Brumbach, "NC Inmate Charged in Kidnapping of DA's Father," Associated Press, April 12, 2014; Lydia Warren and Associated Press, "Bloods Gang Member Who Is Serving Life Sentence 'Masterminded Terrifying Kidnap of Prosecutor's Father Using a Cell Phone He'd Smuggled in to Prison,'" *Daily Mail*, April 11, 2014; Lydia Warren and Associated Press, "Gang Members Who 'Kidnapped Prosecutor's Father and Held Him Captive for Days Had Meant to Capture HER—But They Went to Wrong Address,'" *Daily Mail*, April 23, 2014; Ashley Frantz and AnneClaire Stapleton, "Prosecutor's Dad Kidnapped in 'Elaborate' Plot; FBI Rescues Him," CNN.com, April 10, 2014; Shelley Lynch, "Kidnapping Victim Rescued by FBI Reunited with Family," FBI press release, April 10, 2014, https://www.fbi.gov/charlotte/press-releases/2014/kidnapping-victim-rescued-by-fbi-reunited-with-family; Scott Pelley and Bob Orr, "FBI Told How Its Agents Rescued a North Carolina Man Who Was Kidnapped by Gang Members and Terrorized for Five Days," *CBS Evening News*, April 10, 2014; Marcus K. Garner, "Indictment: Kidnapping Crew Had Wrong Address, Took Wrong Person," *Atlanta Journal Constitution*, April 22, 2014; Andrew Kenney, "Prisoner Charged in Kidnap Conspiracy May Have Had Phone for Weeks," *The Charlotte Observer*, April 11, 2014; "Criminal Complaint Filed Against Kelvin Melton in Kidnapping Case," FBI press release, April 11, 2014, https://www.fbi.gov/charlotte/press-releases/2014/criminal-complaint-filed-against-kelvin-melton-in-kidnapping-case; Colleen Jenkins and Bernadette Baum, "Two More Charged in Gang-Linked Kidnapping of N.C. Prosecutor's Father," Reuters, April 16, 2014; "McDonald's Receipt Leads to Arrest in Wake Forest Kidnapping," *The News and Observer*, April 17, 2014; "Prosecutor—Not Her Father—Was Intended Victim in Wake Forest Kidnapping, Officials Say," *The News and Observer*, April 22, 2014; Patrik Jonsson, "N.C. Prosecutor Kidnap Plot: Home Attacks on Justice Officials on the Upswing," *The Christian Science Monitor*, April 23, 2014; "NC Kidnapping Victim Writes Thank-You Letter," Associated Press, April 29, 2014; Thomas McDonald, "Documents Detail Kidnapping Plot of Wake Prosecutor's Father," *The Charlotte Observer*, July 23, 2014; Daniel Wallis, "Alleged Gangster Admits Lying in North Carolina Kidnap Probe," Reuters, August 29, 2014; Spink John, "FBI Team Rescues a North Carolina Kidnapping Victim," *Atlanta Journal Constitution*, April 11, 2014.

137 Melton's daughters Some observers of the Janssen case have suggested that authorities used a device known as a "stingray," which can identify the precise location of a cellphone, in this investigation. The FBI, when asked about use of a stingray in this case, replied with a response the agency has provided about cell site simulators to other media requests: "Location information is a vital component of law enforcement investigations at the federal, state and local levels. As a general matter, the FBI does not discuss specific techniques used by law enforcement to obtain location information, as they are considered Law Enforcement Sensitive, the public release of which could harm law enforcement efforts at all levels by compromising future use of the technique. The FBI only collects and maintains information that has investigative value and relevance to a case, and such data [are] retained in accordance with controlling federal law and Attorney General policy. The FBI does not keep repositories of cell tower data for any purpose other than in connection with a specific investigation. The collection of cell tower records is only performed after required FBI approvals are received in the specific investigation, and only after the appropriate order is obtained from a court. If the records obtained are deemed relevant, the specific records are made part of the investigative case file. The FBI retains investigative case files in accordance with NARA-approved file retention schedules. If the FBI believes the use of any technology or technique may provide information on an individual where case law dictates that person has a reasonable expectation of privacy, it is FBI policy to obtain a search warrant."

137 directed by Melton himself As noted in the chapter, the details regarding Kelvin Melton, Tianna Brooks (who also allegedly goes by the name Tianna Maynard), and other alleged kidnappers or those allegedly connected to the Janssen kidnapping are contained in court documents or interviews. At the time of writing, Melton, Brooks, and others implicated in this crime have been indicted, but have not gone to trial. Until a trial is conducted and a verdict rendered, allegations remain just that, allegations, and the crimes described in this chapter have not been proven in a court of law. In January 2016, Melton told a court that he was not responsible for the Janssen kidnapping. Other alleged kidnappers are also expected to deny responsibility or guilt. Melton's attorneys, as well as Brooks's attorney, were presented with synopses of all details in this chapter and asked to inquire if their clients, who are incarcerated on other charges or awaiting trial, wished to respond. Brooks's lawyer did not reply. Melton's lawyer, Ryan D. Stump, in an email wrote: "We are under a court order not to discuss the details of Mr. Melton's case and what is contained in the discovery. Unfortunately, due to the restrictions, we are not able to make any comments on the case."

138 predecessors decades before In response to a fact-checking email, a spokeswoman for the FBI said that the bureau's system prior to Sentinel, in addition to using index cards, also used an electronic indexing system. Interviews with agents confirmed this, but said that the electronic system was often incomplete and thus unreliable.

138 rolled out Sentinel In response to a fact-checking email, a spokeswoman for the FBI detailed Sentinel this way: "Sentinel is a tool that manages records; it documents case activities and investigations, the information we own and produce. Sentinel provides a piece of the puzzle. It documents the FBI's work products and is used in conjunction with information we collect or access through other partnerships in order to further data."

138 "agile programming" The words "lean" and "agile" have come to mean different things in different settings. There is, for example, lean product development, lean start-ups, agile management, and agile construction. Some of these definitions or methodologies are very specific. In this chapter, I generally use the phrases in their most global sense. However, for more detailed explanations of the various implementations of these philosophies, I recommend Rachna Shah and Peter T. Ward, "Lean Manufacturing: Context, Practice Bundles, and Performance," *Journal of Operations Management* 21, no. 2 (2003): 129–49; Jeffrey K. Liker, *Becoming Lean: Inside Stories of U.S. Manufacturers* (Portland, Ore.: Productivity Press, 1997); J. Ben Naylor, Mohamed M. Naim, and Danny Berry, "Leagility: Integrating the Lean and Agile Manufacturing Paradigms in the Total Supply Chain," *International Journal of Production Economics* 62, no. 1 (1999): 107–18; Robert Cecil Martin, *Agile Software Development: Principles, Patterns, and Practices* (Upper Saddle River, N.J.: Prentice Hall, 2003); Paul T. Kidd, *Agile Manufacturing: Forging New Frontiers* (Reading, Mass.: Addison-Wesley, 1995); Alistair Cockburn, *Agile Software Development: The Cooperative Game* (Upper Saddle River, N.J.: Addison-Wesley, 2006); Pekka Abrahamsson, Outi Salo, and Jussi Ronkainen, *Agile Software Development Methods: Review and Analysis* (Oulu, Finland: VTT Publications, 2002).

139 "aphrodisiac in Northern California" Rick Madrid passed away in 2012. For my understanding of Mr. Madrid, NUMMI, and General Motors, I am deeply indebted to Frank Langfitt of National Public Radio, Brian Reed of *This American Life,* and other reporters from various newspapers and media organizations who were kind enough to share notes and transcripts with me, as well as Madrid's former colleagues, who shared memories of him. Details on Madrid, including his quotes, draw on a variety of sources, including tapes of interviews with him, notes and transcripts from interviews he gave to other reporters, and recollections of colleagues. In addition, I relied upon Harry Bernstein, "GM Workers Proud of Making the Team," *Los Angeles Times,* June 16, 1987; Clara Germani, "GM-Toyota Venture in California Breaks Tradition, Gets Results," *The Christian Science Monitor,* December 21, 1984; Michelle Levander, "The Divided Workplace: Exhibit Traces Battle for Control of Factory," *Chicago Tribune,* September 17, 1989; Victor F. Zonana, "Auto Venture at Roadblock: GM-Toyota Fremont Plant Produces Happy Workers, High-Quality Product—and a Glut of Unsold Chevrolet Novas," *Los Angeles Times,* December 21, 1987; "NUMMI," *This American Life,* WBEZ Chicago, March 26, 2010; Charles O'Reilly III, "New United Motors Manufacturing, Inc. (NUMMI),"

Stanford Business School Case Studies, no. HR-11, December 2, 1998; Mary-ann Keller, *Rude Awakening: The Rise, Fall, and Struggle for Recovery of General Motors* (New York: William Morrow, 1989); Joel Smith and William Childs, "Imported from America: Cooperative Labor Relations at New United Motor Manufacturing, Inc.," *Industrial Relations Law Journal* (1987): 70–81; John Shook, "How to Change a Culture: Lessons from NUMMI," *MIT Sloan Management Review* 51, no. 2 (2010): 42–51; Michael Maccoby, "Is There a Best Way to Build a Car?" *Harvard Business Review,* November 1997; Daniel Roos, James P. Womack, and Daniel Jones, *The Machine That Changed the World: The Story of Lean Production* (New York: HarperPerennial, 1991); Jon Gertner, "From 0 to 60 to World Domination," *The New York Times,* February 18, 2007; Ceci Connolly, "Toyota Assembly Line Inspires Improvements at Hospital," *The Washington Post,* June 3, 2005; Andrew C. Inkpen, "Learning Through Alliances: General Motors and NUMMI," *Strategic Direction* 22, no. 2 (2006); Paul Adler, "The 'Learning Bureaucracy': New United Motor Manufacturing, Inc." *Research in Organizational Behavior* 15 (1993); "The End of the Line For GM-Toyota Joint Venture," *All Things Considered,* NPR, March 2010; Martin Zimmerman and Ken Basinger, "Toyota Considers Halting Operations at California's Last Car Plant," *Los Angeles Times,* July 24, 2009; Soyoung Kim and Chang-ran Kim, "UPDATE 1—Toyota May Drop U.S. Joint Venture with GM," Reuters, July 10, 2009; Alan Ohnsman and Kae Inoue, "Toyota Will Shut California Plant in First Closure," Bloomberg, August 28, 2009; Jeffrey Liker, *The Toyota Way: 14 Management Principles from the World's Greatest Manufacturer* (New York: McGraw-Hill, 2003); Steven Spear and H. Kent Bowen, "Decoding the DNA of the Toyota Production System," *Harvard Business Review* 77 (1999): 96–108; David Magee, *How Toyota Became #1: Leadership Lessons from the World's Greatest Car Company* (New York: Penguin, 2007).

139　covered his tattoos Keller, *Rude Awakening,* chapter 6.

140　the Fremont plant In a statement sent in response to fact-checking questions, a spokesman for Toyota wrote: "Toyota can't speak to any of the descriptions of the Fremont facility while it operated prior to the independent joint venture with GM. While the broad descriptions of Toyota's philosophy and certain historical facts are consistent with our approach and understanding of events—such as the use of the andon cord, the trip for former GM workers to Japan and the improvement in product quality following the formation of NUMMI—we are unfortunately unable to confirm or provide any other feedback on the specific accounts you provide. However, we can provide the following statement from the company on the NUMMI joint venture, which you are welcome to use if you so choose: 'NUMMI was a groundbreaking model of Japan-U.S. industry collaboration, and we are proud of all its considerable achievements. We remain grateful to all of those involved with NUMMI, including the suppliers, the local community and, most of all, the talented team members who have contributed to the success of this pioneering joint venture.'" In a statement, a spokeswoman for General Motors wrote: "I can't comment on the specific points you shared re the experience at Fremont and NUMMI in the early 1980s, but I can *absolutely* confirm that is not the experience in GM plants today. . . . GM's Global Manufacturing System is a single, common manufacturing system that aligns and engages all employees to use best processes, practices and technologies to eliminate waste throughout the enterprise. . . . While it is true that GMS has its roots in the Toyota Production System (TPS) that was implemented at NUMMI in 1984, many components of GMS grew out of our efforts to benchmark lean manufacturing around the world. . . . While all principles and elements are considered crucial to the successful implementation of GMS, one principle is key to GMS's adaptability, and that is Continuous Improvement. By engaging our employees, we have seen them use GMS to improve our production systems, ensure a safer work environment and improve product quality for our customers."

140   low costs in Japan In a fact-checking email, Jeffrey Liker, who has studied and written extensively about Toyota, wrote: "Toyota realized that to be a global company they needed to set up operations overseas and they had little experience doing it outside of sales. They believed that the Toyota Production System was vital to their success and it was highly dependent on people deeply understanding the philosophy and continuously improving in an environment of trust. They saw NUMMI as a grand experiment to test whether they could make TPS work in the United States with American workers and managers. In fact, in the original agreement with GM they planned on only making Chevy vehicles and when these did not sell because of the negative image of the Chevy brand they brought over the Toyota Corolla. For GM the main attraction was to get some small cars built of good quality profitably and learn how to do this. They seemed to have a passing interest in TPS. For Toyota NUMMI was considered a critical milestone to their future and they studied what was happening every single day to learn as much as they possibly could about operating in the US and developing the Toyota culture overseas."

145   prove their assertion right In response to a fact-checking email, Baron wrote: "Our focus was a bit broader than 'culture.' We were interested in how founders' early choices about organizational design and structuring of employment relationships affected the evolution of their nascent enterprises."

145   "answer a questionnaire" In response to a fact-checking email, Baron wrote that the sources they turned to exceeded just the *San Jose Mercury News:* "We scoured a variety of sources, including the 'Merc,' to try to identify evidence of new foundings. That was supplemented by industry listings from companies like CorpTech (which focuses on marketing targeted to small tech companies). From these sources we put together listings of companies by subsector (biotechnology, semiconductors, etc.). Then we sampled from those listings, seeking to get a representative sampling of firms in terms of age, venture-backed versus not, etc. Somewhat later, after 'the Internet' had emerged as a discernible sector, we replicated the research design focusing specifically on that sector, to see if things were similar or different among the new net companies from the others that we had been studying, and we found the patterns were the same."

146   close to two hundred firms James N. Baron and Michael T. Hannan, "The Economic Sociology of Organizational Entrepreneurship: Lessons from the Stanford Project on Emerging Companies," in *The Economic Sociology of Capitalism,* ed. Victor Nee and Richard Swedberg (New York: Russell Sage, 2002), 168–203; James N. Baron and Michael T. Hannan, "Organizational Blueprints for Success in High-Tech Start-Ups: Lessons from the Stanford Project on Emerging Companies," *Engineering Management Review, IEEE* 31, no. 1 (2003): 16; James N. Baron, M. Diane Burton, and Michael T. Hannan, "The Road Taken: Origins and Evolution of Employment Systems in Emerging Companies," *Articles and Chapters* (1996): 254; James N. Baron, Michael T. Hannan, and M. Diane Burton, "Building the Iron Cage: Determinants of Managerial Intensity in the Early Years of Organizations," *American Sociological Review* 64, no. 4 (1999): 527–47.

146   collected enough data In response to a fact-checking email, Baron wrote: "Perhaps this is nit-picking, but what we were looking at were firms whose founders had similar cultural 'blueprints' or premises underlying their creation. I emphasize this because we were not using observable practices as the basis for differentiation, but instead the way in which founders thought and spoke about their nascent enterprises."

146   one of five categories There were also a sizable number of firms that did not fit neatly into any of the five categories.

147   "on the same path" In response to a fact-checking email, Baron said that he should not be considered an expert on Facebook, and that participants in the study were promised anonymity. He added: "We found that engineering firms fairly frequently evolved, either into bureaucracies or into commitment firms. Those transitions were much less disruptive than others, suggesting that one reason for the popularity of the engineering blueprint at start-up is that it is amenable to being 'morphed' into a different model as the firm matures."

147   " 'You get paid,' " Baron said Baron, in response to a fact-checking email, said that the bureaucratic and the autocratic models have differences but are similar in that "(1) they are both quite infrequent within this sector among start-ups; and (2) they are both unpopular with scientific and technical personnel."

148 successful companies in the world The researchers promised confidentiality to companies that participated in the study, and would not divulge specific firms they had studied.

148 culture came through James N. Baron, Michael T. Hannan, and M. Diane Burton, "Labor Pains: Change in Organizational Models and Employee Turnover in Young, High-Tech Firms," *American Journal of Sociology* 106, no. 4 (2001): 960–1012.

148 *California Management Review* Baron and Hannan, "Organizational Blueprints for Success in High-Tech Start-Ups," 16.

150 "strong advantage" In response to a fact-checking email, Baron expanded upon his comments: "What this doesn't explicitly capture is that commitment firms tended to compete based on superior relationships with their customers over the longer term. It is not just relationships with salespeople, but rather that stable teams of technical personnel, working interdependently with customer-facing personnel, enable these companies to develop technologies that met the needs of their long-term customers."

151 "viability of the Company" Steve Babson, ed., *Lean Work: Empowerment and Exploitation in the Global Auto Industry* (Detroit: Wayne State University Press, 1995).

151 preserve their jobs In a fact-checking email, Jeffrey Liker wrote that Toyota's head of human resources had told a UAW representative that "before laying off any workers they would insource work, then management would take a payout and then they would cut back hours before considering layoffs. In return he said the union needed to agree on three things: 1) competence would be the basis for workers advancing, not seniority, 2) there had to be a minimum of job classifications so they had the flexibility to do multiple jobs, and 3) management and the union would work together on productivity improvements. Within the first year the Chevy Nova was not selling well and they had about 40% too many workers and they kept them all employed in training and doing *kaizen* for several months until they could get the Corolla into production."

154 Harvard researchers wrote Paul S. Adler, "Time-and-Motion Regained," *Harvard Business Review* 71, no. 1 (1993): 97–108.

155 shared power It is important to note that, despite NUMMI's success, the company was not perfect. Its fortunes were tied to the automotive industry, and so when overall car sales declined, NUMMI's profits dipped as well. The NUMMI factory was more expensive to operate than some low-cost foreign competitors, and so there were stretches when the firm was undersold. And when GM tried to export NUMMI's culture to other plants, they found, in some places, it wouldn't take. Enmities between union leaders and managers were simply too deep. Some executives refused to believe that workers, if empowered, would use their authority responsibly. Some employees were unwilling to give GM the benefit of the doubt.

155 "devoted to each other" When the Great Recession hit the automotive industry, NUMMI was one of the casualties. GM, headed toward bankruptcy because of liabilities in other parts of the company, pulled out of the NUMMI partnership in 2009. Toyota concluded it couldn't continue to operate the plant on its own. NUMMI closed in 2010, after manufacturing nearly eight million vehicles.

156 no end in sight Details on development of the Sentinel system come from interviews and Glenn A. Fine, *The Federal Bureau of Investigation's Pre-Acquisition Planning for and Controls over the Sentinel Case Management System*, Audit Report 06-14 (Washington, D.C.: U.S. Department of Justice, Office of the Inspector General, Audit Division, March 2006); Glenn A. Fine, *Sentinel Audit II: Status of the Federal Bureau of Investigation's Case Management System*, Audit Report 07-03 (Washington, D.C.: U.S. Department of Justice, Office of the Inspector General, Audit Division, December 2006); Glenn A. Fine, *Sentinel Audit III: Status of the Federal Bureau of Investigation's Case Management System*, Audit Report 07-40 (Washington, D.C.: U.S. Department of Justice, Office of the Inspector General, Audit Division, August 2007); Raymond J. Beaudet, *Sentinel Audit IV: Status of the Federal Bureau of Investigation's Case Management System*, Audit Report 09-05 (Washington, D.C.: U.S. Department of Justice, Office of the Inspector General, Audit Division, December 2008); Glenn A. Fine, *Sentinel Audit V: Status of the Federal Bureau of Investigation's Case Management System*, Audit Report 10-03 (Washington, D.C.: U.S. Department of Justice, Office of the Inspector General, Audit Division, November 2009); *Status of the Federal Bureau of Investigation's Implementation of the Sentinel Project*, Audit Report 10-22 (Washington, D.C.: U.S. Department of Justice, Office of the Inspector General, March 2010); Thomas J. Harrington, "Response to OIG Report on the FBI's Sentinel Project," FBI press release, October 20, 2010, https://www.fbi.gov/news/pressrel/press-releases/mediaresponse_102010; Cynthia A. Schnedar, *Status of the Federal Bureau of Investigation's Implementation of the Sentinel Project*, Report 12-08 (Washington, D.C.: U.S. Department of Justice, Office of the Inspector General, December 2011); Michael E. Horowitz, *Interim Report on the Federal Bureau of Investigation's Implementation of the Sentinel Project*, Report 12-38 (Washington, D.C.: U.S. Department of Justice, Office of the Inspector General, September 2012);

Michael E. Horowitz, *Audit of the Status of the Federal Bureau of Investigation's Sentinel Program*, Report 14-31 (Washington, D.C.: U.S. Department of Justice, Office of the Inspector General, September 2014); William Anderson et al., *Sentinel Report* (Pittsburgh: Carnegie Mellon Software Engineering Institute, September 2010); David Perera, "Report Questions FBI's Ability to Implement Agile Development for Sentinel," *FierceGovernmentIT,* December 5, 2010, http://www.fiercegovernmentit.com/story/report-questions-fbis-ability -implement-agile-development-sentinel/2010-12-05; David Perera, "FBI: We'll Complete Sentinel with $20 Million and 67 Percent Fewer Workers," *Fierce-GovernmentIT,* October 20, 2010, http://www.fiercegovernmentit.com/story /fbi-well-complete-sentinel-20-million-and-67-percent-fewer-workers/2010 -10-20; Jason Bloomberg, "How the FBI Proves Agile Works for Government Agencies," *CIO,* August 22, 2012, http://www.cio.com/article/2392970/agile -development/how-the-fbi-proves-agile-works-for-government-agencies.html; Eric Lichtblau, "FBI Faces New Setback in Computer Overhaul," *The New York Times,* March 18, 2010; "More Fallout from Failed Attempt to Modernize FBI Computer System," Office of Senator Chuck Grassley, July 21, 2010; "Technology Troubles Plague FBI, Audit Finds," *The Wall Street Journal,* October 20, 2010; "Audit Sees More FBI Computer Woes," *The Wall Street Journal,* October 21, 2010; "FBI Takes Over Sentinel Project," *Information Management Journal* 45, no. 1 (2011); Curt Anderson, "FBI Computer Upgrade Is Delayed," Associated Press, December 23, 2011; Damon Porter, "Years Late and Millions over Budget, FBI's Sentinel Finally On Line," *PC Magazine,* July 31, 2012; Evan Perez, "FBI Files Go Digital, After Years of Delays," *The Wall Street Journal,* August 1, 2012.

156 Toyota Production System philosophy to other industries For more on lean and agile management and methodologies, please see Craig Larman, *Agile and Iterative Development: A Manager's Guide* (Boston: Addison-Wesley Professional, 2004); Barry Boehm and Richard Turner, *Balancing Agility and Discipline: A Guide for the Perplexed* (Boston: Addison-Wesley Professional, 2003); James Shore, *The Art of Agile Development* (Farnham, UK: O'Reilly Media, 2007); David Cohen, Mikael Lindvall, and Patricia Costa, "An Introduction to Agile Methods," *Advances in Computers* 62 (2004): 1–66; Matthias Holweg, "The Genealogy of Lean Production," *Journal of Operations Management* 25, no. 2 (2007): 420–37; John F. Krafcik, "Triumph of the Lean Production System," *MIT Sloan Management Review* 30, no. 1 (1988): 41; Jeffrey Liker and Michael Hoseus, *Toyota Culture: The Heart and Soul of the Toyota Way* (New York: McGraw-Hill, 2007); Steven Spear and H. Kent Bowen, "Decoding the DNA of the Toyota Production System," *Harvard Business Review* 77 (1999): 96–108; James P. Womack and Daniel T. Jones, *Lean Thinking: Banish Waste and Create Wealth in Your Corporation* (New York: Simon & Schuster, 2010);

Stephen A. Ruffa, *Going Lean: How the Best Companies Apply Lean Manufacturing Principles to Shatter Uncertainty, Drive Innovation, and Maximize Profits* (New York: American Management Association, 2008); Julian Page, *Implementing Lean Manufacturing Techniques: Making Your System Lean and Living with It* (Cincinnati: Hanser Gardner, 2004).

156   how software was created "What Is Agile Software Development?" Agile Alliance, June 8, 2013, http://www.agilealliance.org/the-alliance/what-is-agile/; Kent Beck et al., "Manifesto for Agile Software Development," Agile Manifesto, 2001, http://www.agilemanifesto.org/.

157   among many tech firms Dave West et al., "Agile Development: Mainstream Adoption Has Changed Agility," *Forrester Research* 2 (2010): 41.

157   "fix what's broken?" Ed Catmull and Amy Wallace, *Creativity, Inc.: Overcoming the Unseen Forces That Stand in the Way of True Inspiration* (New York: Random House, 2014).

157   wrote in 2005 J. P. Womack and D. Miller, *Going Lean in Health Care* (Cambridge, Mass.: Institute for Healthcare Improvement, 2005).

158   get Sentinel working Jeff Stein, "FBI Sentinel Project Is over Budget and Behind Schedule, Say IT Auditors," *The Washington Post*, October 20, 2010.

158   plan everything in advance This method of planning is often known as a "waterfall approach," because it is a sequential design methodology in which progress "flows" downward from conception to initiation, analysis, design, construction, testing, production/implementation, and maintenance. At the core of this approach is the belief that each stage can be anticipated and scheduled.

158   unfettered themselves In response to a fact-checking email, Fulgham expanded his comments: "I assigned the CTO (Jeff Johnson) as the day to day executive for oversight. We hired an Agile Scrum Master (Mark Crandall) to serve as a coach and mentor (not as a project manager). We created an open physical workspace in the basement that allowed collaborative communications between team members. We assigned three Cyber Special Agents as the front end development leads, and the Director, Deputy Director and I empowered them to recommend any process improvements and/or form consolidations (in order not to just digitize any potentially outdated processes/forms). I worked with the CEOs of our top vendors for the products that were going to make up Sentinel to get their support and their best cleared personnel. The team adopted (under Mark's coaching) the agile methodology. All FBI stakeholders were part of the business side of the Sentinel team to ensure their needs were met. The technical team conducted self directed two-week sprints. We had nightly automated builds. A dedicated QA team was located with the development team, and I held a meeting every two weeks to view fully functional code (no mockups) and personally signed off on requirements. All stakeholders, the DOJ, the DOJ IG, the White House and other interested government agencies, attended these demo days to observe our progress and process."

160   solve thousands of crimes In response to a fact-checking email, a spokeswoman for the FBI wrote, regarding Sentinel: "We are not predicting crime. We may identify trends and threats."

160   "is capable of" Jeff Sutherland, *Scrum: The Art of Doing Twice the Work in Half the Time* (New York: Crown Business, 2014).

162   "cultural mindset" Robert S. Mueller III, "Statement Before the House Permanent Select Committee on Intelligence," Washington, D.C., October 6, 2011, https://www.fbi.gov/news/testimony/the-state-of-intelligence-reform -10-years-after-911.

CHAPTER SIX: DECISION MAKING

167   worth $450,000 Throughout this chapter, chips are referred to by their notional dollar value. However, it is important to note that in tournaments like this one, chips are tokens that are collected to determine winners—they are not traded in for cash on a one-to-one basis. Rather, prize money is paid out based on how someone places in the competition. So someone could have $200,000 in chips and take fifth place in a tournament and win $300,000, for instance. In this particular tournament, the prize was $2 million and, by coincidence, the total number of chips was also $2 million.

167   prize for second place The 2004 Tournament of Champions is described in slightly different chronological order than what occurred in order to highlight the salient points of each hand. Beyond describing hands out of order, no other facts have been changed. For my understanding of the 2004 Tournament of Champions as well as poker more generally, I am indebted to Annie Duke, Howard Lederer, and Phil Hellmuth for their time and advice. In addition, this account relies upon the taped version of the 2004 TOC, provided by ESPN; Annie Duke, with David Diamond, *How I Raised, Folded, Bluffed, Flirted, Cursed and Won Millions at the World Series of Poker* (New York: Hudson Street Press, 2005); "Annie Duke: The Big Things You Don't Do," *The Moth Radio Hour*, September 13, 2012, http://themoth.org/posts /stories/the-big-things-you-dont-do; "Annie Duke: A House Divided," *The Moth Radio Hour*, July 20, 2011, http://themoth.org/posts/stories/a-house -divided; "Dealing with Doubt," *Radiolab*, season 11, episode 4, http://www

.radiolab.org/story/278173-dealing-doubt/; Dina Cheney, "Flouting Convention, Part II: Annie Duke Finds Her Place at the Poker Table," *Columbia College Today,* July 2004, http://www.college.columbia.edu/cct_archive/jul04/features4.php; Ginia Bellafante, "Dealt a Bad Hand? Fold 'Em. Then Raise," *The New York Times,* January 19, 2006; Chuck Darrow, "Annie Duke, Flush with Success," *The Philadelphia Inquirer,* June 8, 2010; Jamie Berger, "Annie Duke, Poker Pro," *Columbia Magazine,* March 4, 2013, http://www.columbia.edu/cu/alumni/Magazine/Spring2002/Duke.html; "Annie Duke Profile," *The Huffington Post,* February 21, 2013; Del Jones, "Know Yourself, Know Your Rival," *USA Today,* July 20, 2009; Richard Deitsch, "Q&A with Annie Duke," *Sports Illustrated,* May 26, 2005; Mark Sauer, "Annie Duke Found Her Calling," *San Diego Union-Tribune,* October 9, 2005; George Sturgis Coffin, *Secrets of Winning Poker* (Wilshire, 1949); Richard D. Harroch and Lou Krieger, *Poker for Dummies* (New York: Wiley, 2010); David Sklansky, *The Theory of Poker* (Two Plus Two Publishers, 1999); Michael Bowling et al., "Heads-Up Limit Hold'em Poker Is Solved," *Science* 347, no. 6218 (2015): 145–49; Darse Billings et al., "The Challenge of Poker," *Artificial Intelligence* 134, no. 1 (2002): 201–40; Kevin B. Korb, Ann E. Nicholson, and Nathalie Jitnah, "Bayesian Poker," *Proceedings of the Fifteenth Conference on Uncertainty in Artificial Intelligence* (San Francisco: Morgan Kaufmann, 1999).

167 she was going to win Gerald Hanks, "Poker Math and Probability," *Pokerology,* http://www.pokerology.com/lessons/math-and-probability/.

170 win a Nobel Prize Daniel Kahneman and Amos Tversky, "Prospect Theory: An Analysis of Decision Under Risk," *Econometrica: Journal of the Econometric Society* 47, no. 2 (1979): 263–91.

174 a million television viewers The tournament drew an estimated 1.5 million viewers.

174 She's not sure Annie, in a phone call to check facts in this chapter, expanded upon her thinking: "If Greg had jacks or better, I was in a bad situation. I was very undecided about the hand he could be holding, and I was in a situation where I really did have to create more certainty for myself. I really needed to decide if he had aces or kings, and then fold. Also, Greg Raymer, at that point, was an unknown quantity, but my brother and I had been watching videotapes of him play, and we had seen what we thought was a "tell," something he did physically when he had a good hand, and I saw him do this particular thing that suggested to me that he had a strong hand. That's not a certain thing, you don't know if a tell is 100 percent, but it helped tip me into thinking he had a strong hand."

175 "intelligence forecasts" "Aggregative Contingent Estimation," Office of the Director of National Intelligence (IARPA), 2014, Web.

176　some fresh ideas For my understanding of the Good Judgment Project, I am indebted to Barbara Mellers et al., "Psychological Strategies for Winning a Geopolitical Forecasting Tournament," *Psychological Science* 25, no. 5 (2014): 1106–15; Daniel Kahneman, "How to Win at Forecasting: A Conversation with Philip Tetlock," *Edge,* December 6, 2012, https://edge.org/conversation/how-to-win-at-forecasting; Michael D. Lee, Mark Steyvers, and Brent Miller, "A Cognitive Model for Aggregating People's Rankings," *PloS One* 9, no. 5 (2014); Lyle Ungar et al., "The Good Judgment Project: A Large Scale Test" (2012); Philip Tetlock, *Expert Political Judgment: How Good Is It? How Can We Know?* (Princeton, N.J.: Princeton University Press, 2005); Jonathan Baron et al., "Two Reasons to Make Aggregated Probability Forecasts More Extreme," *Decision Analysis* 11, no. 2 (2014): 133–45; Philip E. Tetlock et al., "Forecasting Tournaments Tools for Increasing Transparency and Improving the Quality of Debate," *Current Directions in Psychological Science* 23, no. 4 (2014): 290–95; David Ignatius, "More Chatter than Needed," *The Washington Post,* November 1, 2013; Alex Madrigal, "How to Get Better at Predicting the Future," *The Atlantic,* December 11, 2012; Warnaar et al., "Aggregative Contingent Estimation System"; Uriel Haran, Ilana Ritov, and Barbara A. Mellers, "The Role of Actively Open-Minded Thinking in Information Acquisition, Accuracy, and Calibration," *Judgment and Decision Making* 8, no. 3 (2013): 188–201; David Brooks, "Forecasting Fox," *The New York Times,* March 21, 2013; Philip Tetlock and Dan Gardner, *Seeing Further* (New York: Random House, 2015).

176　A group of At various points during the GJP, the precise number of researchers involved fluctuated.

176　questions as the experts In response to a fact-checking email, Barbara Mellers and Philip Tetlock, another of the GJP leaders, wrote: "We had two different types of training in the first year of the tournament. One was probabilistic reasoning and the other was scenario training. Probabilistic reasoning worked somewhat better, so in subsequent years, we implemented only the probabilistic training. Training was revised each year. As it evolved, there was a section on geopolitical reasoning and another on probabilistic reasoning. . . . Here is a section that describes the training: We constructed educational modules on probabilistic-reasoning training and scenario training that drew on state-of-the-art recommendations. Scenario training taught forecasters to generate new futures, actively entertain more possibilities, use decision trees, and avoid biases such as over-predicting change, creating incoherent scenarios, or assigning probabilities to mutually exclusive and exhaustive outcomes that exceed 1.0. Probability training guided forecasters to consider reference classes, average multiple estimates from existing models, polls, and expert panels, extrapolate over time when variables were continuous, and avoid judgmental traps such as overconfidence, the confirmation bias, and base-rate neglect.

Each training module was interactive with questions and answers to check participant understanding."

177   abilities to forecast the future In response to a fact-checking email, Don Moore wrote: "On average, those with training did better. But not everyone who got trained did better than all the people who did not get it."

180   "tremendously useful" Brooks, "Forecasting Fox."

181   "things you aren't sure about" In response to a fact-checking email, Don Moore wrote: "What makes our forecasters good is not just their high level of accuracy, but their well-calibrated humility. They are no more confident than they deserve to be. It's ideal to know when you have forecast the future with accuracy and when you haven't."

183   or roughly 20 percent In an email, Howard Lederer, a two-time World Series of Poker champion, explained the further nuances required in analyzing this hand: "The hand you use as an example is MUCH more complicated than it appears." Given what's known, Lederer said, there is actually a better than 20 percent chance of winning. "Here's why. If you KNOW your opponent has an A or a K, then you know seven cards. Your two [cards], your opponent's one card, and the four [communal cards] on the board. This means there are 45 unknown cards (you have no information on your opponent's other card). This would mean you have nine hearts to win, and 36 non-hearts to lose. The odds would be 4 to 1, or 1 in 5. The percentages are 20%. As long as you are not putting more than 20% of the money into the pot, it's a good call. Here's where you might ask: if I am only 20% to win against an A or K, then how can I be better than [20%] to win? Your opponent might not have an A or K! He could have a spade flush draw without an A or K, he could have a straight draw with a 5–6. He could have a lower heart draw. That would be great for you! There's also a chance he just has garbage and is trying to bluff you with nothing. In general, I'd calculate the chances that your opponent has one of these drawing or bluffing hands at about 30% (given how many of these possibilities there are). So let's do some probabilistic math: 70% of the time he has an A or K, and you win 20% of those times. 25% of the time he has a draw and you win about 82% of those hands (I'm combining various possible odds given his range of holdings when he is drawing). And 5% of the time he has a total bluff and you win 89% of the time when he has garbage. Your total chances of winning are: $(.7 \times .2) + (.25 \times .82) + (.05 \times .89) = 39\%$! This is a simple 'expected value' calculation. You can see that the .7, .25 and .05 part of the calculation adds up to 1. Meaning we have covered all the possible holdings and assigned them probabilities. And we are making our best guess as to our chances against each holding. At the table, you don't have time to do all the math, but 'in your gut' you can feel the odds and make the easy call.

One other note, if you miss your flush and your opponent bets, you should seriously consider calling anyway. You will be getting well over 10–1, and the chances he is bluffing are probably higher than that. This is just a simple taste of the complexity of poker."

183 they'll quit For more on calculating odds in poker, please see Pat Dittmar, *Practical Poker Math: Basic Odds and Probabilities for Hold'em and Omaha* (Toronto: ECW Press, 2008); "Poker Odds for Dummies," CardsChat, https://www.cardschat.com/odds-for-dummies.php; Kyle Siler, "Social and Psychological Challenges of Poker," *Journal of Gambling Studies* 26, no. 3 (2010): 401–20.

185 "odds work for you" In response to a fact-checking email, Howard Lederer wrote: "It's more complex than that. Amateurs players make many different kinds of errors. Some play too loose. They crave the uncertainty and favor action over prudence. Some players are too conservative, favoring a small loss in a hand over taking the chance to win, but also the chance to take a large loss. Your job as a poker pro is to simply play your best each hand. In the long run, your superior decisions will defeat your opponent's poor decisions, whatever they may be. The societal value of poker is that it is a great training ground for learning sound decision-making under conditions of uncertainty. Once you get the hang of playing poker, you develop the skills necessary to make probabilistic decisions in life."

185 Annie's brother, Howard Though it does not bear on the events described in this chapter, disclosure compels mentioning that Lederer was a founder and board member of Tiltware, LLC, the company behind Full Tilt Poker, a popular website that was accused of bank fraud and illegal gambling by the U.S. Department of Justice. In 2012, Lederer settled a civil lawsuit with the Department of Justice related to Full Tilt Poker. He admitted no wrongdoing, but did agree to forfeit more than $2.5 million.

186 winning this hand Technically, Howard has an 81.5 percent chance of winning—however, because it is hard to win half a hand of poker, this has been rounded up to 82 percent.

187 remaining cards on the table In response to a fact-checking email, Howard Lederer wrote: "I would say that in a 3 handed situation, [a pair of sevens] is close to 90% to be best before the flop. This is the hand where I agree anyone would have played her hand and my hand the same way; all in before the flop. After we had all the money in, I am not a slight favorite, but instead a large favorite. This [is] a unique feature of hold'em. If you have a slightly better hand than your opponent, you are often a big favorite. 7–7 is about 81% to beat 6–6."

188 "they tell you might occur" In response to a fact-checking email, Howard Lederer wrote: "It's not an easy thing to choose a profession where you lose more often than you win. One has to focus on the long run, and realize that if you get offered 10–1, on enough 5–1 shots, you will come out ahead, while also realizing that you will lose 5 out of 6 times."

188 humans process information Tenenbaum, in an email responding to fact-checking questions, described his research this way: "Often we start with what looks like a gap between humans and computers, where humans are outperforming standard computers with intuitions that may not look like computations. . . . But then we try to close that gap, by understanding how human intuitions actually have a subtle computational basis, which then can be engineered in a machine, to make the machine smarter in more human-like ways."

189 "seeing just a few examples" Joshua B. Tenenbaum et al., "How to Grow a Mind: Statistics, Structure, and Abstraction," *Science* 331, no. 6022 (2011): 1279–85.

189 "examples of each?" Ibid.

191 (which has no strong pattern) In an email responding to fact-checking questions, Tenenbaum said that many of the examples they used were fairly complex, and "the reasons for the prediction functions having these shapes are the combination of (1) the priors, plus (2) a certain assumption about when an event is likely to be sampled (the 'likelihood'), (3) Bayesian updating from priors to posteriors, and (4) using the 50th percentile of the posterior as the basis for prediction. What's correct about what you have is that in our simple model, only (1) varies across domains—between movies, representatives, life spans, etc.—while (2–4) are the same for all the tasks. But [it's] because of these causal processes (which vary across domains) together with the rest of the statistical computations (which are the same across domains) that the prediction functions have the shape they do." It is important to note that the graphs in this text do not represent accurate empirical results, but rather patterns of predictions—the estimations that represent the 50th percentile of being right or wrong.

191 You read about a movie These are summaries of the questions asked. The direct wording of each question was: "Imagine you hear about a movie that has taken in 60 million dollars at the box office, but don't know how long it has been running. What would you predict for the total amount of box office intake for that movie?" "Insurance agencies employ actuaries to make predictions about people's life spans—the age at which they will die—based upon demographic information. If you were assessing an insurance case for a 39-year-old man, what would you predict for his life span?" "Imagine you are in somebody's kitchen and notice that a cake is in the oven. The timer shows that it has been baking for 14 minutes. What would you predict for the total amount of time the cake needs to bake?" "If you heard a member of the House of Representatives had served for 11 years, what would you predict his total term in the House would be?"

192 variation of Bayes' rule In an email responding to fact-checking questions, Tenenbaum wrote that "the most natural way to make these kinds of predictions in computers is to run algorithms which effectively implement the logic of Bayes' rule. The computers typically don't explicitly 'use' Bayes' rule, because the direct computations of Bayes' rule are typically intractable to carry out except in simple cases. Rather the programmers give the computers prediction algorithms whose predictions are made to be approximately consistent with Bayes' rule in a wide range of cases, including these."

193 data and your assumptions Sheldon M. Ross, *Introduction to Probability and Statistics for Engineers and Scientists* (San Diego: Academic Press, 2004).

195 skewed, as well "Base rate" typically refers to a yes-or-no question. In the Tenenbaum experiment, participants were asked to make numerical predictions, rather than answer a binary question, and so it's most accurate to refer to this assumption as a "prior distribution."

195 failures we've overlooked In an email responding to fact-checking questions, Tenenbaum wrote that "It's not clear from our work that predictions for events in a certain class improve progressively with more experience with events of that type. Sometimes they might, sometimes they don't. And this is not the only way to acquire a prior. As the pharaohs example shows, and other projects by us and other researchers, people can acquire a prior in various ways beyond direct experience with a class of events, including being told things, making analogies to other classes of events, forming analogies, and so on."

197 "the Poker Brat" Eugene Kim, "Why Silicon Valley's Elites Are Obsessed with Poker," *Business Insider,* November 22, 2014, http://www.businessinsider.com/best-poker-players-in-silicon-valley-2014-11.

198 "bluff when it matters" In response to a fact-checking email, Hellmuth wrote: "Annie is a great poker player, and she has stood the test of time. I respect her, and I respect her Hold'em game."

199 He folds In response to a fact-checking email, Hellmuth wrote: "I think she was trying to tilt me (get me emotional and upset) by showing a nine in that situation. A lot of players would have gone broke with my hand there (top pair) w[ith] a 'Safe' turn card, but I've made a living deviating from the norm and trusting my instincts (my white magic, my reading ability). I trusted it and folded."

201 middle of the table In response to a fact-checking email, Hellmuth wrote: "With the chips I had at that time I had to go all in w[ith] 10–8 on that flop (I had top pair and there were flush draws, and straight draws possible). Completely standard. If you're trying to imply that I put the money because I was emotionally tilted, you're wrong. Nothing I could do there."

202   Phil is out In response to a fact-checking email, Hellmuth contends that he and Annie had struck a deal when the tournament came down to the two of them in which they pledged to guarantee each other $750,000 regardless of the winner, and play for the last $500,000. Annie Duke confirmed this deal.

CHAPTER SEVEN: INNOVATION

205   movie everyone is talking about For my understanding of *Frozen*'s development, I am particularly indebted to Ed Catmull, Jennifer Lee, Andrew Millstein, Peter Del Vecho, Kristen Anderson-Lopez, Bobby Lopez, Amy Wallace, and Amy Astley, as well as other Disney employees, some of whom wished to remain anonymous, who were generous with their time. Additionally, I relied upon Charles Solomon, *The Art of Frozen* (San Francisco: Chronicle Books, 2015); John August, "*Frozen* with Jennifer Lee," *Scriptnotes*, January 28, 2014, http://johnaugust.com/2014/frozen-with-jennifer-lee; Nicole Laporte, "How *Frozen* Director Jennifer Lee Reinvented the Story of the Snow Queen," *Fast Company*, February 28, 2014; Lucinda Everett, "*Frozen:* Inside Disney's Billion-Dollar Social Media Hit," *The Telegraph*, March 31, 2014; Jennifer Lee, "*Frozen*, Final Shooting Draft," Walt Disney Animation Studios, September 23, 2013, http://gointothestory.blcklst.com/wp-content/uploads/2014/11/Frozen .pdf; "*Frozen:* Songwriters Kristen Anderson-Lopez and Robert Lopez Official Movie Interview," YouTube, October 31, 2013, https://www.youtube.com /watch?v=mzZ77n4Ab5E; Susan Wloszczyna, "With *Frozen*, Director Jennifer Lee Breaks Ice for Women Directors," *Indiewire*, November 26, 2013, http:// blogs.indiewire.com/womenandhollywood/with-frozen-director-jennifer-lee-breaks-the-ice-for-women-directors; Jim Hill, "Countdown to Disney *Frozen:* How One Simple Suggestion Broke the Ice on the Snow Queen's Decades-Long Story Problems," *Jim Hill Media*, October 18, 2013, http://jimhillmedia .com/editor_in_chief1/b/jim_hill/archive/2013/10/18/countdown-to-disney -quot-frozen-quot-how-one-simple-suggestion-broke-the-ice-on-the-quot -snow-queen-quot-s-decades-long-story-problems.aspx; Brendon Connelly, "Inside the Research, Design, and Animation of Walt Disney's *Frozen* with Producer Peter Del Vecho," *Bleeding Cool*, September 25, 2013, http://www .bleedingcool.com/2013/09/25/inside-the-research-design-and-animation -of-walt-disneys-frozen-with-producer-peter-del-vecho/; Ed Catmull and Amy Wallace, *Creativity, Inc.: Overcoming the Unseen Forces That Stand in the Way of True Inspiration* (New York: Random House, 2014); Mike P. Williams,

"Chris Buck Reveals True Inspiration Behind Disney's *Frozen* (Exclusive),"
Yahoo! Movies, April 8, 2014; Williams College, "Exploring the Songs of *Frozen* with Kristen Anderson-Lopez '94," YouTube, June 30, 2014, https://www
.youtube.com/watch?v=ftddAzabQMM; Dan Sarto, "Directors Chris Buck and
Jennifer Lee Talk *Frozen*," Animation World Network, November 7, 2013; Jennifer Lee, "Oscars 2014: *Frozen*'s Jennifer Lee on Being a Female Director," *Los
Angeles Times*, March 1, 2014; Rob Lowman, "Unfreezing *Frozen*: The Making
of the Newest Fairy Tale in 3D by Disney," *Los Angeles Daily News*, November 19, 2013; Jill Stewart, "Jennifer Lee: Disney's New Animation Queen,"
LA Weekly, May 15, 2013; Simon Brew, "A Spoiler-Y, Slightly Nerdy Interview About Disney's *Frozen*," *Den of Geek!*, December 12, 2013, http://www
.denofgeek.com/movies/frozen/28567/a-spoiler-y-nerdy-interview-about-disneys-frozen; Sean Flynn, "Is It Her Time to Shine?" *The Newport Daily News*,
February 17, 2014; Mark Harrison, "Chris Buck and Jennifer Lee Interview:
On Making *Frozen*," *Den of Geek!* December 6, 2013, http://www.denofgeek
.com/movies/frozen/28495/chris-buck-and-jennifer-lee-interview-on-making
-frozen; Mike Fleming, "Jennifer Lee to Co-Direct Disney Animated Film *Frozen*," *Deadline Hollywood*, November 29, 2012; Rebecca Keegan, "Disney Is
Reanimated with *Frozen*, *Big Hero 6*," *Los Angeles Times*, May 9, 2013; Lindsay
Miller, "On the Job with Jennifer Lee, Director of *Frozen*," *Popsugar*, February 28, 2014, http://www.popsugar.com/celebrity/Frozen-Director-Jennifer
-Lee-Interview-Women-Film-33515997; Trevor Hogg, "Snowed Under: Chris
Buck Talks About Frozen," *Flickering Myth*, March 26, 2014, http://www
.flickeringmyth.com/2014/03/snowed-under-chris-buck-talks-about.html;
Jim Hill, "Countdown to Disney *Frozen*: The Flaky Design Idea Behind the
Look of Elsa's Ice Palace," *Jim Hill Media*, October 9, 2013, http://jimhillmedia
.com/editor_in_chief1/b/jim_hill/archive/2013/10/09/countdown-to-disney
-quot-frozen-quot-the-flaky-design-idea-behind-the-look-of-elsa-s-ice-palace
.aspx; Rebecca Keegan, "Husband-Wife Songwriting Team's Emotions Flow
in *Frozen*," *Los Angeles Times*, November 1, 2013; Heather Wood Rudulph,
"Get That Life: How I Co-Wrote the Music and Lyrics for *Frozen*," *Cosmopolitan*, April 27, 2015; Simon Brew, "Jennifer Lee and Chris Buck Interview:
Frozen, Statham, *Frozen 2*," *Den of Geek!*, April 4, 2014, http://www.denofgeek
.com/movies/frozen/29346/jennifer-lee-chris-buck-interview-frozen
-statham-frozen-2; Carolyn Giardina, "Oscar: With *Frozen*, Disney Invents a
New Princess," *The Hollywood Reporter*, November 27, 2013; Steve Persall,
"Review: Disney's *Frozen* Has a Few Cracks in the Ice," *Tampa Bay Times*,
November 26, 2013; Kate Muir, "Jennifer Lee on Her Disney Hit *Frozen*: We
Wanted the Princess to Kick Ass," *The Times*, December 12, 2013; "Out of the
Cold," *The Mail on Sunday*, December 29, 2013; Kathryn Shattuck, "*Frozen*

Directors Take Divide-and-Conquer Approach," *The New York Times*, January 16, 2014; Ma'ayan Rosenzweig and Greg Atria, "The Story of *Frozen:* Making a Disney Animated Classic," *ABC News Special Report*, September 2, 2014, http://abcnews.go.com/Entertainment/fullpage/story-frozen-making-disney -animated-classic-movie-25150046; Amy Edmondson et al., "Case Study: Teaming at Disney Animation," *Harvard Business Review*, August 27, 2014.

207 surprised by all the criticisms In an email sent in response to fact-checking questions, Andrew Millstein, president of Disney Animation Studios, wrote: "These are the kind of notes that fuel our creative process and help propel the forward progress of all of our films in production. The creative leadership on any film often gets too close to their films and loses objectivity. Our Story Trust functions like a highly critical and skilled audience that can point to flaws in the story-telling and, more important, provide potential solutions. . . . You're describing a process of experimentation, exploration and discovery that are key components of all our films. It's not a question of if this will happen, but to what degree. This is a constant part of our process and the expectation [of] every filmmaking team. It is what contributes to the high standards that our films set."

207 *Book of Mormon* In an email sent in response to fact-checking questions, Bobby Lopez made clear that Kristen was a sounding board for him in writing *Avenue Q* and *Book of Mormon* but was not formally credited on those shows.

208 dozens of others popped up In an email sent in response to fact-checking questions, a spokeswoman for Walt Disney Animation Studios wrote that the studio wished to emphasize "how typical this process is for every film at Disney Animation since John [Lasseter] and Ed [Catmull] have become our studio leaders—the screening process, the notes sessions, the taking apart of the film and putting it back together. This is typical, not atypical."

209 "good ideas are suffocated" In an email sent in response to fact-checking questions, Ed Catmull, president of Disney Animation, wrote that the various anecdotes in this chapter are "viewpoints of different snapshots in time as the film developed. . . . In truth, you could substitute different words and it would pretty much describe how *every* film goes through searching and change. This is worth emphasizing so that people don't have the impression that *Frozen* was different in that way."

209 *Frozen* was winding down In an email sent in response to fact-checking questions, Millstein wrote: "Creativity needs time, space and support to fully explore multiple ideas simultaneously. Our creative leadership has to have the confidence and trust in each other to experiment, fail and try again and again until the answers to story questions and problems get better and more refined. There also needs to be a relentless focus on finding the best solutions to difficult and thorny problems and never settling for sub-optimum solutions because of time issues. Our creative teams need to trust that the executive management fundamentally believes in and supports this process."

209 avant-garde on Broadway Amanda Vaill, *Somewhere: The Life of Jerome Robbins* (New York: Broadway Books, 2008); "Q&A with Producer Director Judy Kinberg, 'Jerome Robbins: Something to Dance About,'" directed by Judy Kinberg, *American Masters*, PBS, January 28, 2009, http://www.pbs.org/wnet/americanmasters/jerome-robbins-q-a-with-producerdirector-judy-kinberg/1100/; Sanjay Roy, "Step-by-Step Guide to Dance: Jerome Robbins," *The Guardian*, July 7, 2009; Sarah Fishko, "The Real Life Drama Behind West Side Story," NPR, January 7, 2009, http://www.npr.org/2011/02/24/97274711/the-real-life-drama-behind-west-side-story; Jeff Lundun and Scott Simon, "Part One: Making a New Kind of Musical," NPR, September 26, 2007, http://www.npr.org/templates/story/story.php?storyId=14730899; Jeff Lundun and Scott Simon, "Part Two: Casting Calls and Out of Town Trials," NPR, September 26, 2007, http://www.npr.org/templates/story/story.php?storyId=14744266; Jeff Lundun and Scott Simon, "Part Three: Broadway to Hollywood—and Beyond," NPR, September 26, 2007, http://www.npr.org/templates/story/story.php?storyId=14749729; "West Side Story Film Still Pretty, and Witty, at 50," NPR, October 17, 2011, http://www.npr.org/2011/10/17/141427333/west-side-story-still-pretty-and-witty-at-50; Jesse Green, "When You're a Shark You're a Shark All the Way," *New York Magazine*, March 15, 2009; Larry Stempel, "The Musical Play Expands," *American Music* 10, no. 2 (1992): 136–69; Beth Genné, "'Freedom Incarnate': Jerome Robbins, Gene Kelly, and the Dancing Sailors as an Icon of American Values in World War II," *Dance Chronicle* 24, no. 1 (2001): 83–103; Bill Fischer and Andy Boynton, "Virtuoso Teams," *Harvard Business Review*, July 1, 2005; Otis L. Guernsey, ed., *Broadway Song and Story: Playwrights/Lyricists/Composers Discuss Their Hits* (New York: Dodd Mead, 1985); Larry Stempel, *Showtime: A History of the Broadway Musical Theater* (New York: W. W. Norton, 2010); Robert Emmet Long, "West Side Story," in *Broadway, the Golden Years: Jerome Robbins and the Great Choreographer-Directors: 1940 to the Present* (New York: Continuum, 2001); Leonard Bernstein, "A West Side Log" (1982); Terri Roberts, "West Side Story: 'We Were All Very Young,'" *The Sondheim Review* 9, no. 3 (Winter 2003); Steven Suskin, *Opening Night on Broadway: A Critical Quotebook of the Golden Era of the Musical Theatre, Oklahoma! (1943) to Fiddler on the Roof (1964)* (New York: Schirmer Trade Books, 1990); Amanda Vaill, "Jerome Robbins—About the Artist," *American Masters*, PBS, January 27, 2009, http://www.pbs.org/wnet/americanmasters/jerome-robbins-about-the-artist/1099/.

210   **actor on the stage** There are a few outliers to this musical formula, most notably *Oklahoma!*, in which dance was used to express plot and emotional moments.

210   **"me a ballet?"** Tim Carter, "Leonard Bernstein: West Side Story. By Nigel Simeone," *Music and Letters* 92, no. 3 (2011): 508–10.

210   **would be *West Side Story*** *West Side Story* went through numerous names before the final title was chosen.

211   **musical's main characters** Excerpts of letters come from the Leonard Bernstein Collection at the Library of Congress as well as from records made available by various authors and the New York Public Library system.

211   **"jitterbugging"** This was written by Leonard Bernstein, as quoted in *The Leonard Bernstein Letters* (New Haven, Conn.: Yale University Press, 2013).

211   **"we're boring the audience"** Jerome Robbins, as quoted in *The Leonard Bernstein Letters* (New Haven, Conn.: Yale University Press, 2013).

211   **"two intermissions"** Vaill, *Somewhere*.

211   **"Shakespeare standing behind you"** Ibid.

211   **"Forget Anita"** Deborah Jowitt, *Jerome Robbins: His Life, His Theater, His Dance* (New York: Simon & Schuster, 2004).

212   ***Science* in 2013** Brian Uzzi et al., "Atypical Combinations and Scientific Impact," *Science* 342, no. 25 (2013): 468–72.

212   **Brian Uzzi and Ben Jones** For more on Uzzi and Jones's work, please see Stefan Wuchty, Benjamin F. Jones, and Brian Uzzi, "The Increasing Dominance of Teams in Production of Knowledge," *Science* 316, no. 5827 (2007): 1036–39; Benjamin F. Jones, Stefan Wuchty, and Brian Uzzi, "Multi-University Research Teams: Shifting Impact, Geography, and Stratification in Science," *Science* 322, no. 5905 (2008): 1259–62; Holly J. Falk-Krzesinski et al., "Advancing the Science of Team Science," *Clinical and Translational Science* 3, no. 5 (2010): 263–66; Ginger Zhe Jin et al., *The Reverse Matthew Effect: Catastrophe and Consequence in Scientific Teams* (working paper 19489, National Bureau of Economic Research, 2013); Brian Uzzi and Jarrett Spiro, "Do Small Worlds Make Big Differences? Artist Networks and the Success of Broadway Musicals, 1945–1989" (unpublished manuscript, Evanston, Ill., 2003); Brian Uzzi, and Jarrett Spiro, "Collaboration and Creativity: The Small World Problem," *American Journal of Sociology* 111, no. 2 (2005): 447–504; Brian Uzzi, "A Social Network's Changing Statistical Properties and the Quality of Human Innovation," *Journal of Physics A: Mathematical and Theoretical* 41, no. 22 (2008); Brian Uzzi, Luis A.N. Amaral, and Felix Reed-Tsochas, "Small-World Networks and Management Science Research: A Review," *European Management Review* 4, no. 2 (2007): 77–91.

214 creative and important In response to a fact-checking email, Uzzi wrote: "The other thing is that teams are more likely to get this sweet spot of creativity right. They are more likely than individuals to put together atypical combinations of prior sources. Also, a paper with the right mix of conventional and atypical ideas by a team does better than a single author, given the same mix of conventional and atypical ideas. This means teams are better than individuals at sourcing and deriving insights from atypical combinations."

214 bought lottery tickets Amos Tversky and Daniel Kahneman, "Availability: A Heuristic for Judging Frequency and Probability," *Cognitive Psychology* 5, no. 2 (1973): 207–32; Daniel Kahneman and Amos Tversky, "Prospect Theory: An Analysis of Decision Under Risk," *Econometrica: Journal of the Econometric Society* 47, no. 2 (1979): 263–91; Amos Tversky and Daniel Kahneman, "Judgment Under Uncertainty: Heuristics and Biases," *Science* 185, no. 4157 (1974): 1124–31; Amos Tversky and Daniel Kahneman, "The Framing of Decisions and the Psychology of Choice," *Science* 211, no. 4481 (1981): 453–58; Daniel Kahneman and Amos Tversky, "Choices, Values, and Frames," *American Psychologist* 39, no. 4 (1984): 341; Daniel Kahneman, *Thinking, Fast and Slow* (New York: Farrar, Straus and Giroux, 2011); Daniel Kahneman and Amos Tversky, "On the Psychology of Prediction," *Psychological Review* 80, no. 4 (1973): 237.

214 how genes evolve Qiong Wang et al., "Naive Bayesian Classifier for Rapid Assignment of rRNA Sequences into the New Bacterial Taxonomy," *Applied and Environmental Microbiology* 73, no. 16 (2007): 5261–67; Jun S. Liu, "The Collapsed Gibbs Sampler in Bayesian Computations with Applications to a Gene Regulation Problem," *Journal of the American Statistical Association* 89, no. 427 (1994): 958–66.

214 "railway and mining" Andrew Hargadon and Robert I. Sutton, "Technology Brokering and Innovation in a Product Development Firm," *Administrative Science Quarterly* 42, no. 4 (1997): 716–49.

215 gambling techniques René Carmona et al., *Numerical Methods in Finance: Bordeaux, June 2010*, Springer Proceedings in Mathematics, vol. 12 (Berlin: Springer Berlin Heidelberg, 2012); René Carmona et al., "An Introduction to Particle Methods with Financial Application," in *Numerical Methods in Finance*, 3–49; Pierre Del Moral, *Mean Field Simulation for Monte Carlo Integration* (Boca Raton, Fla.: CRC Press, 2013); Roger Eckhardt, "Stan Ulam, John von Neumann, and the Monte Carlo Method," *Los Alamos Science*, special issue (1987): 131–37.

215 in the shape of a hat Andrew Hargadon and Robert I. Sutton, "Technology Brokering and Innovation in a Product Development Firm," *Administrative Science Quarterly* 42, no. 4 (1997): 716–49; Roger P. Brown, "Polymers in Sport and Leisure," *Rapra Review Reports* 12, no. 3 (November 2, 2001); Melissa Larson, "From Bombers to Bikes," *Quality* 37, no. 9 (1998): 30.

215   child-rearing techniques Benjamin Spock, *The Common Sense Book of Baby and Child Care* (New York: Pocket Books, 1946).

215   "evaluated as valuable" Ronald S. Burt, "Structural Holes and Good Ideas," *American Journal of Sociology* 110, no. 2 (2004): 349–99.

215   succeeded somewhere else In an email sent in response to fact-checking questions, Burt wrote: "Managers offered their best idea for improving the value of their function to the company. The two senior executives in the function evaluated each idea (stripped of personal identification). The summary evaluation of each idea turned out to be primarily predicted by the extent to which the person who articulated the idea had a network that reached across boundaries (structural holes) between network groups, functions, divisions in the company."

215   pushed the right way For more on the concept of brokerage, please see Ronald S. Burt, *Structural Holes: The Social Structure of Competition* (Cambridge, Mass.: Harvard University Press, 2009); Ronald S. Burt, "The Contingent Value of Social Capital," *Administrative Science Quarterly* 42, no. 2 (1997): 339–65; Ronald S. Burt, "The Network Structure of Social Capital," in B. M. Staw and R. I. Sutton, *Research in Organizational Behavior*, vol. 22 (New York: Elsevier Science JAI, 2000), 345–423; Ronald S. Burt, *Brokerage and Closure: An Introduction to Social Capital* (New York: Oxford University Press, 2005); Ronald S. Burt, "The Social Structure of Competition," *Explorations in Economic Sociology* 65 (1993): 103; Lee Fleming, Santiago Mingo, and David Chen, "Collaborative Brokerage, Generative Creativity, and Creative Success," *Administrative Science Quarterly* 52, no. 3 (2007): 443–75; Satu Parjanen, Vesa Harmaakorpi, and Tapani Frantsi, "Collective Creativity and Brokerage Functions in Heavily Cross-Disciplined Innovation Processes," *Interdisciplinary Journal of Information, Knowledge, and Management* 5, no. 1 (2010): 1–21; Thomas Heinze and Gerrit Bauer, "Characterizing Creative Scientists in Nano-S&T: Productivity, Multidisciplinarity, and Network Brokerage in a Longitudinal Perspective," *Scientometrics* 70, no. 3 (2007): 811–30; Markus Baer, "The Strength-of-Weak-Ties Perspective on Creativity: A Comprehensive Examination and Extension," *Journal of Applied Psychology* 95, no. 3 (2010): 592; Ajay Mehra, Martin Kilduff, and Daniel J. Brass, "The Social Networks of High and Low Self-Monitors: Implications for Workplace Performance," *Administrative Science Quarterly* 46, no. 1 (2001): 121–46.

216   plot's central tensions I am indebted to the New York Public Library for making an early draft version of the *West Side Story* script available to me. This is an abridgment of that script, shortened for ease of representation.

219 communicated through dance This text is a combination of finished versions of the *West Side Story* script, Robbins's notes, and interviews providing a description of the choreography from the first staging of the show and other sources.

220 "essential dramatic information" Larry Stempel, "The Musical Play Expands," *American Music* (1992): 136–69.

220 the original Maria Fishko, "Real Life Drama Behind *West Side Story*."

221 coffee cups and to-do lists The *Frozen* core team included Buck, Lee, Del Vecho, Bobby Lopez and Kristen Anderson-Lopez, Paul Briggs, Jessica Julius, Tom MacDougall, Chris Montan, and, at times, others from various departments.

222 upstate New York In an email sent in response to fact-checking questions, a spokeswoman for Walt Disney Animation Studios wrote that Lee "and her sister fought, as kids do; they grew together as they grew older. They were never estranged. . . . In college, they became close. They lived together in NYC for a while, even."

223 "ourselves on the screen" In an email sent in response to fact-checking questions, Millstein wrote: "Solutions to story issues [are often] connected to personal emotional experiences. We draw from our own stories, history and emotional lives as a wellspring of inspiration. . . . We also draw on the experiences of others throughout the studio and deep research into specific areas that a film may attempt to explore. In the case of *Frozen*, we had a built-in research group at Disney Animation: employees who are sisters. They can describe firsthand what it's like to have a sister as a sibling and the life experiences they've had. This is wonderful firsthand source material."

223 "their experiences than other people" Gary Wolf, "Steve Jobs: The Next Insanely Great Thing," *Wired*, April 1996.

223 "pushed to use it sometimes" In an email sent in response to fact-checking questions, Catmull wrote: "It is too simple to say that people need to be pushed. Yes, they do, but they also need to be allowed to create, and we must make it safe for them to find something new. Andrew and I both need to be a force to make things move along, while at the same time, trying to keep fear from slowing them down or getting stuck. This is what makes the job so hard."

224 make them stay put Art Fry, "The Post-it note: An Intrapreneurial Success," *SAM Advanced Management Journal* 52, no. 3 (1987): 4.

224 from wine spills P. R. Cowley, "The Experience Curve and History of the Cellophane Business," *Long Range Planning* 18, no. 6 (1985): 84–90.

224 middle of the night Lewis A. Barness, "History of Infant Feeding Practices," *The American Journal of Clinical Nutrition* 46, no. 1 (1987): 168–70; Donna A. Dowling, "Lessons from the Past: A Brief History of the Influence of Social, Economic, and Scientific Factors on Infant Feeding," *Newborn and Infant Nursing Reviews* 5, no. 1 (2005): 2–9.

224 psychologist Gary Klein Gary Klein, *Seeing What Others Don't: The Remarkable Ways We Gain Insights* (New York: PublicAffairs, 2013).

226 of people's expectations In an email sent in response to fact-checking questions, Bobby Lopez wrote: "From our perspective—we hit 'send' on an email with our mp3 attached, and then count the minutes, hours, or sometimes days before we hear back from them. Sometimes it means something and sometimes it doesn't. We didn't hear back right away, so we began to doubt the song, but when they did call us it was clear they were very excited."

226 "feel like one of us" In an email sent in response to fact-checking questions, a spokeswoman for Walt Disney Animation Studios wrote that Lee "had written a draft of the script already in April [2012 in] which Elsa was a more sympathetic character but there was still a plan for her to turn evil halfway through the film. ["Let It Go"] first appeared in [an] August 2012 screening. "Let It Go" helped shift the tone of the Elsa character. It should be noted that John Lasseter felt a personal tie to this as well—when thinking of Elsa, he thought of his son, Sam, and his juvenile diabetes. When Sam was getting poked and prodded as a child, he turned to John and said, 'Why me?' It wasn't Sam's fault he had diabetes, just as it is not Elsa's fault she has these icy powers."

226 "It had to feel real" In an email sent in response to fact-checking questions, a spokeswoman for Walt Disney Animation Studios wrote that Chris Buck had a vision for how the film would end. "The ending—making it work emotionally[—]was a puzzle. By October 2012, Jennifer had the ending envisioning the four main characters in a blizzard of fear, which story artist John Ripa boarded. Ripa's boards received a standing ovation from John Lasseter in the room. As Jennifer says, 'We knew the end, we just needed to earn it.'"

227 pressure that comes from deadlines Teresa M. Amabile et al., "Assessing the Work Environment for Creativity," *Academy of Management Journal* 39, no. 5 (1996): 1154–84; Teresa M. Amabile, Constance N. Hadley, and Steven J. Kramer, "Creativity Under the Gun," *Harvard Business Review* 80, no. 8 (2002): 52–61; Teresa M. Amabile, "How to Kill Creativity," Harvard Business Review 76, no. 5 (1998): 76–87; Teresa M. Amabile, "A Model of Creativity and Innovation in Organizations," *Research in Organizational Behavior* 10, no. 1 (1988): 123–67.

227 **"Lee a second director"** In an email sent in response to fact-checking questions, Catmull wrote that it is important to emphasize that Lee was a second director, not a "codirector," which has multiple meanings in Hollywood. "There is an actual title of 'Co-director' which is at a lower level than 'director.' At Disney we frequently have two directors who both have the title of 'director.' In this case, both Jenn and Chris were equal directors. . . . Jenn was made director along with Chris."

228 **spinning in place** In an email sent in response to fact-checking questions, Millstein wrote: "Jenn's promotion to an equal directing partner with Chris provided an opportunity to alter the team dynamics in a positive way and their receptivity to potential new ideas. . . . Jenn is a very sensitive and emphatic filmmaker. Her sensitivity to team dynamics, her role and voice and deep need to maintain a deep collaboration is what helped make *Frozen* successful." One additional factor influencing the decision to promote Lee to director, according to Buck, was that at that time, one of his children had a health issue that required attention, and as a result "John and Ed and Andrew saw my personal need, and they asked me, right before, what would you think of having Jenn as a co-director? And I said yes, I said absolutely, I would love that."

228 **ecologically bland** I am indebted to the help of Stephen Palumbi of Stanford's Hopkins Marine Station and Elizabeth Alter of the City University of New York for their assistance in my understanding of the intermittent disturbance hypothesis.

228 **distributed so unevenly** Joseph H. Connell, "Diversity in Tropical Rain Forests and Coral Reefs," *Science*, n.s. 199, no. 4335 (1978): 1302–10.

230 **intermediate disturbance hypothesis** Like many scientific theories, the intermediate disturbance hypothesis has many parents. For a more complete history, please see David M. Wilkinson, "The Disturbing History of Intermediate Disturbance," *Oikos* 84, no. 1 (1999): 145–47.

230 **"nor too frequent"** John Roth and Mark Zacharias, *Marine Conservation Ecology* (London: Routledge, 2011).

230 **staple of biology** For more on the intermediate disturbance hypothesis, including the perspectives of those who challenge the theory, please see Wilkinson, "The Disturbing History of Intermediate Disturbance"; Jane A. Catford et al., "The Intermediate Disturbance Hypothesis and Plant Invasions: Implications for Species Richness and Management," *Perspectives in Plant Ecology, Evolution and Systematics* 14, no. 3 (2012): 231–41; John Vandermeer et al., "A Theory of Disturbance and Species Diversity: Evidence from Nicaragua After Hurricane Joan," *Biotropica* 28, no. 4 (1996): 600–613; Jeremy

W. Fox, "The Intermediate Disturbance Hypothesis Should Be Abandoned," *Trends in Ecology and Evolution* 28, no. 2 (2013): 86–92.

233 Lee sat down with John Lasseter In an email sent in response to fact-checking questions, Catmull wrote that figuring out *Frozen*'s ending was a team effort. John Ripa, an animator at Disney, storyboarded the ending. "This was a powerful and influential part of the development of the story. . . . [In addition] there was a particularly impactful offsite where a great deal of progress was made."

234 "tell the team," said Lasseter In an email sent in response to fact-checking questions, a spokeswoman for Walt Disney Animation Studios wrote: "Jennifer feels this is very, very important: This was a story Jennifer and Chris did together. This was a partnership. [The emails] Kristen shared were based on conversations Jennifer was having with Chris daily. Chris is just as much a part of these conversations as Jennifer, Kristen and Bobby. . . . This is [Chris Buck's] film, first and foremost."

CHAPTER EIGHT: ABSORBING DATA

238 multiplication quiz "Dante Williams" is a pseudonym used to protect the privacy of a student who was a minor when these events occurred.

239 "Peace Bowl" Ben Fischer, "Slaying Halts 'Peace Bowl,'" *Cincinnati Enquirer*, August 13, 2007.

240 guide Cincinnati's efforts Marie Bienkowski et al., *Enhancing Teaching and Learning Through Educational Data Mining and Learning Analytics: An Issue Brief* (Washington, D.C.: U.S. Department of Education, Office of Technology, October 2012), https://tech.ed.gov/wp-content/uploads/2014/03/edm-la-brief.pdf.

240 "we were on board" For more on Elizabeth Holtzapple's research and Cincinnati Public Schools' approach to data usage, I recommend Elizabeth Holtzapple, "Criterion-Related Validity Evidence for a Standards-Based Teacher Evaluation System," *Journal of Personnel Evaluation in Education* 17, no. 3 (2003): 207–19; Elizabeth Holtzapple, *Report on the Validation of Teachers Evaluation System Instructional Domain Ratings* (Cincinnati: Cincinnati Public Schools, 2001).

240 basic educational benchmarks "South Avondale Elementary: Transformation Model," Ohio Department of Education, n.d.

240 the "Elementary Initiative" Information on the EI and other Cincinnati Public Schools reforms came from various sources, including Kim McGuire, "In Cincinnati, They're Closing the Achievement Gap," *Star Tribune*, May 11,

2004; Alyson Klein, "Education Week, Veteran Educator Turns Around Cincinnati Schools," *Education Week*, February 4, 2013; Nolan Rosenkrans, "Cincinnati Offers Toledo Schools a Road Map to Success," *The Blade*, May 13, 2012; Gregg Anrig, "How to Turn an Urban School District Around—Without Cheating," *The Atlantic*, May 9, 2013; John Kania and Mark Kramer, "Collective Impact," *Stanford Social Innovation Review* 9, no. 1 (Winter 2011): 36–41; Lauren Morando Rhim, *Learning How to Dance in the Queen City: Cincinnati Public Schools' Turnaround Initiative*, Darden/Curry Partnership for Leaders in Education (Charlottesville: University of Virginia, 2011); Emily Ayscue Hassel and Bryan C. Hassel, "The Big U Turn," *Education Next* 9, no. 1 (2009): 20–27; Rebecca Herman et al., *Turning Around Chronically Low-Performing Schools: A Practice Guide* (Washington, D.C.: National Center for Education Evaluation and Regional Assistance, Institute of Education Sciences, U.S. Department of Education, 2008); *Guide to Understanding Ohio's Accountability System, 2008–2009* (Columbus: Ohio Department of Education, 2009), Web; Daniela Doyle and Lyria Boast, *2010 Annual Report: The University of Virginia School Turnaround Specialist Program*, Darden/Curry Partnership for Leaders in Education, Public Impact (Charlottesville: University of Virginia, 2011); Dana Brinson et al., *School Turnarounds: Actions and Results*, Public Impact (Lincoln, Ill.: Center on Innovation and Improvement, 2008); L. M. Rhim and S. Redding, eds., *The State Role in Turnaround: Emerging Best Practices* (San Francisco: WestEd, 2014); William S. Robinson and LeAnn M. Buntrock, "Turnaround Necessities," *The School Administrator* 68, no. 3 (March 2011): 22–27; Susan McLester, "Turnaround Principals," *District Administration* (May 2011); Daniel Player and Veronica Katz, "School Improvement in Ohio and Missouri: An Evaluation of the School Turnaround Specialist Program" (CEPWC Working Paper Series no. 10, University of Virginia, Curry School of Education, June 2013), Web; Alison Damast, "Getting Principals to Think Like Managers," *Bloomberg Businessweek*, February 16, 2012; "CPS 'Turnaround Schools' Lift District Performance," *The Cincinnati Herald*, August 21, 2010; Dakari Aarons, "Schools Innovate to Keep Students on Graduation Track," *Education Week*, June 2, 2010; "Facts at a Glance," Columbia Public Schools K–12, n.d., Web.

241   how to *use* it The Cincinnati Public School system's Elementary Initiative had other components in addition to instructing teachers in how to use data. Those included using data and analysis to guide evidence-based decisions; implementing a new principal evaluation system aligned to the district's strategic plan that included student performance scores; expanding school-site learning teams of teachers to build capacity in all schools; training primary and intermediate content specialists in core subjects; and becoming more family friendly and community engaged. "Using data and evidence, we will improve practice, differentiate instruction, and track learning results for every student," the district wrote in a summary of the initiative. "Our goal is to create a collaborative learning culture that involves families, is embraced in schools and is supported by the Board, central office and the community. Such a culture is at the heart of the elementary school initiative. . . . Just as the medical community uses diagnostics to determine treatment for critical care patients, so are we using data and analysis with 15 critical care schools to reshape training, support and delivery of services aligned to the academic, social and emotional needs of the students." ("Elementary Initiative: Ready for High School," Cincinnati Public Schools, 2014, http://www.cps-k12.org /academics/district-initiatives/elementary-initiative.) It is also worth noting that, though everyone spoken to in reporting this chapter credits a data-driven approach with fueling South Avondale's transformation, they also noted that such changes were possible only because of strong leadership at the school and commitment from teachers.

241 **inner-city reform** "Elementary Initiative: Ready for High School."

241 **state math exam** Ibid.; South Avondale Elementary School Rank-ing," School Digger, 2014, http://www.schooldigger.com/go/OH/schools /0437500379/school.aspx; "South Avondale Elementary School Profile," Great Schools, 2013, Web.

241 **the school district read** "School Improvement, Building Profiles, South Avondale," Ohio Department of Education, 2014, Web.

242 **but more useful** For more on the role of data in classroom improve-ment, please see Thomas J. Kane et al., "Identifying Effective Classroom Prac-tices Using Student Achievement Data," *Journal of Human Resources* 46, no 3 (2011): 587–613; Pam Grossman et al., "Measure for Measure: A Pilot Study Linking English Language Arts Instruction and Teachers' Value-Added to Stu-dent Achievement" (CALDER Working Paper no. 45, Calder Urban Institute, May 2010); Morgaen L. Donaldson, "So Long, Lake Wobegon? Using Teacher Evaluation to Raise Teacher Quality," Center for American Progress, June 25, 2009, Web; Eric Hanushek, "Teacher Characteristics and Gains in Stu-dent Achievement: Estimation Using Micro-Data," *The American Economic Review* 61, no. 2 (1971): 280–88; Elizabeth Holtzapple, "Criterion-Related Va-lidity Evidence for a Standards-Based Teacher Evaluation System," *Journal of Personnel Evaluation in Education* 17, no. 3 (2003): 207–19; Brian A. Jacob and Lars Lefgren, *Principals as Agents: Subjective Performance Measurement in Education* (working paper no. w11463, National Bureau of Economic Re-search, 2005); Brian A. Jacob, Lars Lefgren, and David Sims, *The Persistence of Teacher-Induced Learning Gains* (working paper no. w14065, National Bureau of Economic Research, 2008); Thomas J. Kane and Douglas O. Staiger, *Esti-*

mating Teacher Impacts on Student Achievement: An Experimental Evaluation (working paper no. w14607, National Bureau of Economic Research, 2008); Anthony Milanowski, "The Relationship Between Teacher Performance Evaluation Scores and Student Achievement: Evidence from Cincinnati," *Peabody Journal of Education* 79, no. 4 (2004): 33–53; Richard J. Murnane and Barbara R. Phillips, "What Do Effective Teachers of Inner-City Children Have in Common?" *Social Science Research* 10, no. 1 (1981): 83–100; Steven G. Rivkin, Eric A. Hanushek, and John F. Kain, "Teachers, Schools, and Academic Achievement," *Econometrica* 73, no. 2 (2005): 417–58.

243 less stressful Jessica L. Buck, Elizabeth McInnis, and Casey Randolph, *The New Frontier of Education: The Impact of Smartphone Technology in the Classroom,* American Society for Engineering Education, 2013 ASEE Southeast Section Conference; Neal Lathia et al., "Smartphones for Large-Scale Behavior Change Interventions," *IEEE Pervasive Computing* 3 (2013): 66–73; "Sites That Help You Track Your Spending and Saving," *Money Counts: Young Adults and Financial Literacy,* NPR, May 18, 2011; Shafiq Qaadri, "Meet a Doctor Who Uses a Digital Health Tracker and Thinks You Should Too," *The Globe and Mail,* September 4, 2014; Claire Cain Miller, "Collecting Data on a Good Night's Sleep," *The New York Times,* March 10, 2014; Steven Beasley and Annie Conway, "Digital Media in Everyday Life: A Snapshot of Devices, Behaviors, and Attitudes," Museum of Science and Industry, Chicago, 2011; Adam Tanner, "The Web Cookie Is Dying. Here's the Creepier Technology That Comes Next," *Forbes,* June 17, 2013, http://www.forbes.com /sites/adamtanner/2013/06/17/the-web-cookie-is-dying-heres-the-creepier -technology-that-comes-next/.

243 harder to decide For more on information overload and information blindness, please see Martin J. Eppler and Jeanne Mengis, "The Concept of Information Overload: A Review of Literature from Organization Science, Accounting, Marketing, MIS, and Related Disciplines," *The Information Society* 20, no. 5 (2004): 325–44; Pamela Karr-Wisniewski and Ying Lu, "When More Is Too Much: Operationalizing Technology Overload and Exploring Its Impact on Knowledge Worker Productivity," *Computers in Human Behavior* 26, no. 5 (2010): 1061–72; Joseph M. Kayany, "Information Overload and Information Myths," Itera, n.d., http://www.itera.org/wordpress/wp-content /uploads/2012/09/ITERA12_Paper15.pdf; Marta Sinclair and Neal M. Ashkanasy, "Intuition Myth or a Decision-Making Tool?" *Management Learning* 36, no. 3 (2005): 353–70.

243 blanket of powder Snow blindness can also refer to a burn of the cornea, which is the front surface of the eye, by ultraviolet B rays.

243 enroll in 401(k) plans Sheena S. Iyengar, Gur Huberman, and Wei Jiang, "How Much Choice Is Too Much? Contributions to 401(k) Retirement

Plans," *Pension Design and Structure: New Lessons from Behavioral Finance* (Philadelphia: Pension Research Council, 2004): 83–95.

244 more than thirty plans In an email sent in response to fact-checking questions, Tucker Kuman, a colleague of the paper's lead author, Sheena Sethi-Iyengar, wrote: "What was observed in the analysis was that, everything else being equal, every ten funds added was associated with a 1.5 percent to 2 percent drop in employee participation rate (peak participation—75%—occurred when 2 funds were offered). . . . As the offerings increased in number, the decline in participation rates is exacerbated. If you look at the graphic representation [Figure 5–2 in the paper] of the relationship between participation and number of funds offered, you'll notice we begin to see a *steeper* decline in participation rates when the number of funds hits about 31."

244 information overload Jeanne Mengis and Martin J. Eppler, "Seeing Versus Arguing the Moderating Role of Collaborative Visualization in Team Knowledge Integration," *Journal of Universal Knowledge Management* 1, no. 3 (2006): 151–62; Martin J. Eppler and Jeanne Mengis, "The Concept of Information Overload: A Review of Literature from Organization Science, Accounting, Marketing, MIS, and Related Disciplines," *The Information Society* 20, no. 5 (2004): 325–44.

245 "winnowing" or "scaffolding" Fergus I. M. Craik and Endel Tulving, "Depth of Processing and the Retention of Words in Episodic Memory," *Journal of Experimental Psychology: General* 104, no. 3 (1975): 268; Monique Ernst and Martin P. Paulus, "Neurobiology of Decision Making: A Selective Review from a Neurocognitive and Clinical Perspective," *Biological Psychiatry* 58, no. 8 (2005): 597–604; Ming Hsu et al., "Neural Systems Responding to Degrees of Uncertainty in Human Decision-Making," *Science* 310, no. 5754 (2005): 1680–83.

245 hardly aware it's occurring For more on the decision-making aspect of scaffolding and cognition, please see Gerd Gigerenzer and Wolfgang Gaissmaier, "Heuristic Decision Making," *Annual Review of Psychology* 62 (2011): 451–82; Laurence T. Maloney, Julia Trommershäuser, and Michael S. Landy, "Questions Without Words: A Comparison Between Decision Making Under Risk and Movement Planning Under Risk," *Integrated Models of Cognitive Systems* (2007): 297–313; Wayne Winston, *Decision Making Under Uncertainty* (Ithaca, N.Y.: Palisade Corporation, 1999); Eric J. Johnson and Elke U. Weber, "Mindful Judgment and Decision Making," *Annual Review of Psychology* 60 (2009): 53; Kai Pata, Erno Lehtinen, and Tago Sarapuu, "Inter-Relations of Tutor's and Peers' Scaffolding and Decision-Making Discourse Acts," *Instructional Science* 34, no. 4 (2006): 313–41; Priscilla Wohlstetter, Amanda Datnow, and Vicki Park, "Creating a System for Data-Driven Decision Making: Applying the Principal-Agent Framework," *School Effectiveness and School Improve-*

ment 19, no. 3 (2008): 239–59; Penelope L. Peterson and Michelle A. Comeaux, "Teachers' Schemata for Classroom Events: The Mental Scaffolding of Teachers' Thinking During Classroom Instruction," *Teaching and Teacher Education* 3, no. 4 (1987): 319–31; Darrell A. Worthy et al., "With Age Comes Wisdom: Decision Making in Younger and Older Adults," *Psychological Science* 22, no. 11 (2011): 1375–80; Pat Croskerry, "Cognitive Forcing Strategies in Clinical Decisionmaking," *Annals of Emergency Medicine* 41, no. 1 (2003): 110–20; Brian J. Reiser, "Scaffolding Complex Learning: The Mechanisms of Structuring and Problematizing Student Work," *The Journal of the Learning Sciences* 13, no. 3 (2004): 273–304; Robert Clowes and Anthony F. Morse, "Scaffolding Cognition with Words," in *Proceedings of the Fifth International Workshop on Epigenetic Robotics: Modeling Cognitive Development in Robotic Systems* (Lund, Sweden: Lund University Cognitive Studies, 2005), 101–5.

246 make a choice For more on disfluency, please see Adam L. Alter, "The Benefits of Cognitive Disfluency," *Current Directions in Psychological Science* 22, no. 6 (2013): 437–42; Adam L. Alter et al., "Overcoming Intuition: Metacognitive Difficulty Activates Analytic Reasoning," *Journal of Experimental Psychology: General* 136, no. 4 (2007): 569; Adam L. Alter, *Drunk Tank Pink: And Other Unexpected Forces That Shape How We Think, Feel, and Behave* (New York: Penguin, 2013); Adam L. Alter et al., "Overcoming Intuition: Metacognitive Difficulty Activates Analytic Reasoning," *Journal of Experimental Psychology: General* 136, no. 4 (2007): 569; Adam L. Alter and Daniel M. Oppenheimer, "Effects of Fluency on Psychological Distance and Mental Construal (or Why New York Is a Large City, but New York Is a Civilized Jungle)," *Psychological Science* 19, no. 2 (2008): 161–67; Adam L. Alter and Daniel M. Oppenheimer, "Uniting the Tribes of Fluency to Form a Metacognitive Nation," *Personality and Social Psychology Review* 13, no. 3 (2009): 219–35; John Hattie and Gregory C. R. Yates, *Visible Learning and the Science of How We Learn* (London: Routledge, 2013); Nassim Nicholas Taleb, *Antifragile: Things That Gain from Disorder* (New York: Random House, 2012); Daniel M. Oppenheimer, "The Secret Life of Fluency," *Trends in Cognitive Sciences* 12, no. 6 (2008): 237–41; Edward T. Cokely and Colleen M. Kelley, "Cognitive Abilities and Superior Decision Making Under Risk: A Protocol Analysis and Process Model Evaluation," *Judgment and Decision Making* 4, no. 1 (2009): 20–33; Connor Diemand-Yauman, Daniel M. Oppenheimer, and Erikka B. Vaughan, "Fortune Favors the Bold (and the Italicized): Effects of Disfluency on Educational Outcomes," *Cognition* 118, no. 1 (2011): 111–15; Hyunjin Song and Norbert Schwarz, "Fluency and the Detection of Misleading Questions: Low Processing Fluency Attenuates the Moses Illusion," *Social Cognition* 26, no. 6 (2008): 791–99; Anuj K. Shah and Daniel M. Oppenheimer, "Easy Does It: The Role of Fluency in Cue Weighting," *Judgment and Decision Making* 2, no. 6

(2007): 371–79. In an email sent in response to fact-checking questions, Adam Alter, a professor at NYU who has studied disfluency, explained disfluency as "the sense of mental difficulty that people experience when they try to process (make sense of) certain pieces of information—complex words; text printed in ornate fonts; text printed against background of a similar color; drawing dimly remembered ideas from memory; struggling to remember a phone number; etc. You don't have to be manipulating or using data, per se, for an experience to be disfluent. Some of this turns on how you define data—it sounds like you're defining it very broadly, so perhaps your definition comes close to mine if you think of every cognitive process as 'using data.' "

246 **easier to digest** Alter wrote in an email that some recent work "challenges the disfluency literature. . . . Some of my friends/colleagues have written another piece ["Disfluent Fonts Don't Help People Solve Math Problems"] that shows how finicky the effect is; [and] how hard it can be to replicate at least one of the effects (the cognitive reflection test effects)."

247 **"using it in conversations"** In an email sent in response to fact-checking questions, Adam Alter expanded on his quote to note that disfluency causes learning to be "longer lasting, perhaps, but certainly deeper. We don't comment much on decay rates—how long the information is retained—but it probably follows that ideas last longer when they're processed more deeply. . . . The more they elaborate on that information, the more they tend to remember it. That's a general principle from cognitive psychology. If I ask you to remember the word 'balloon,' you'll remember it more easily if, at the point of storing it in memory, you imagine a red balloon floating into the sky, or you think of a baboon carrying a balloon, or you otherwise do more than just trying to cram the word into your already overstuffed memory bank."

247 **pay their credit card bills** Chase Manhattan Bank, now known as JPMorgan Chase, was provided with a summary of all facts contained in this chapter. A representative for the company wrote: "Given that more than 15 years have passed [since] the merger of Bank One and J. P. Morgan Chase in 2004, it's been difficult to find the right internal sources for this."

251 **"pick up on things"** In an email sent in response to fact-checking questions, Fludd wrote that there were other elements to her management style that she believes contributed to her success: "I also was able to identify that the collectors had different learning styles that caused them to interpret the data in different ways that could either negatively or positively impact their performance. . . . Management would accuse me of spoiling my collectors because sometimes I would cook them breakfast on the weekends. Food always helped. Being a minister often helped me relate to the collectors and assist them in ways that other managers couldn't. I would visit family members in the hospital, perform marriages, prayer requests. Collectors knew I was a no nonsense manager, but they also knew I cared about them. . . . Knowing how to interpret data and explaining it in a way that is meaningful and relevant is important. The collectors having access to data that was relevant to their performance was important. However if you cannot give the employee a road map on how to take the data that they are receiving and show them how to get to their desired performance destination, then it means nothing. How you relay that data is just as important. The important thing every manager needs to remember is not to forget the human side of the data they are relating."

252 various experiments In an email sent in response to fact-checking questions, Niko Cantor wrote: "It is also true that Charlotte was a better manager than most peers, more engaging, more enrolling her people in a quest to become better. She did make the job feel more like a game. I think some of the effects of the collectors listening better and therefore connecting better because the collectors were more engaged were important."

253 "So *you're* Ms. Johnson" Johnson started her teaching career at Pleasant Hill Elementary, and then later joined South Avondale, serving as a teacher coach.

255 over the PA system The "Hot Pencil Drills" were unique to South Avondale, and not done in all of the schools participating in the Elementary Initiative.

256 Delia Morris was a "Delia Morris" is a pseudonym used to protect the privacy of a student who was a minor when these events occurred.

258 "the engineering design process" Yousef Haik and Tamer Shahin, *Engineering Design Process* (Independence, Ky.: Cengage Learning, 2010); Clive L. Dym et al., *Engineering Design: A Project-Based Introduction* (New York: Wiley, 2004); Atila Ertas and Jesse C. Jones, *The Engineering Design Process* (New York: Wiley, 1996); Thomas J. Howard, Stephen J. Culley, and Elies Dekoninck, "Describing the Creative Design Process by the Integration of Engineering Design and Cognitive Psychology Literature," *Design Studies* 29, no. 2 (2008): 160–80.

258 teacher's manual explained "What is the Engineering Design Process?" Innovation First International, http://curriculum.vexrobotics.com/curriculum /intro-to-engineering/what-is-the-engineering-design-process.

261 their own experiences Stephen J. Hoch, "Availability and Interference in Predictive Judgment," *Journal of Experimental Psychology: Learning, Memory, and Cognition* 10, no. 4 (1984): 649.

261 question was framed In an email sent in response to fact-checking questions, the author of this study, Stephen Hoch, wrote: "The only other thing that I might add is that old ideas can get in the way of new ideas, creating interference and essentially blocking the thought process. One way to overcome the interference is to take a break so that the old ideas die down in terms of their salience."

261 hard to dislodge Irwin P. Levin, Sandra L. Schneider, and Gary J. Gaeth, "All Frames Are Not Created Equal: A Typology and Critical Analysis of Framing Effects," *Organizational Behavior and Human Decision Processes* 76, no. 2 (1998): 149–88; Hilary A. Llewellyn-Thomas, M. June McGreal, and Elaine C. Thiel, "Cancer Patients' Decision Making and Trial-Entry Preferences: The Effects of 'Framing' Information About Short-Term Toxicity and Long-Term Survival," *Medical Decision Making* 15, no. 1 (1995): 4–12; David E. Bell, Howard Raiffa, and Amos Tversky, *Decision Making: Descriptive, Normative, and Prescriptive Interactions* (Cambridge: Cambridge University Press, 1988); Amos Tversky and Daniel Kahneman, "Rational Choice and the Framing of Decisions," *The Journal of Business* 59, no. 4, part 2 (1986): S251–78.

262 "inside their heads" In response to a fact-checking email, Johnson wrote: "The idea is that we think of a subset of the relevant information."

264 program named "Gen-1" Lekan Oguntoyinbo, "Hall Sweet Home," *Diverse Issues in Higher Education* 27, no. 25 (2011): 8; Dana Jennings, "Second Home for First Gens," *The New York Times*, July 20, 2009.

265 the difference between students Pam A. Mueller and Daniel M. Oppenheimer, "The Pen Is Mightier Than the Keyboard: Advantages of Longhand over Laptop Note Taking," *Psychological Science* 25, no. 6 (2014).

265 verbatim phrases In a note sent in response to fact-checking questions, the first author of this study, Pam Mueller of Princeton, wrote: "Only because a lot of people (on the Internet) seem to assume that we didn't randomly assign participants to groups, and therefore the conclusions are invalid, it might be worth mentioning that the two groups were, in fact, randomly assigned. We did ask students about their underlying note-taking preference, but due to small numbers of participants in certain conditions (e.g., longhand-preferring students at Princeton assigned to the laptop condition) we can't draw strong conclusions about any interactions there. There is some suggestion that those who preferred longhand in their regular note taking were more effective than others when using a laptop (i.e., continuing to take shorter, non-verbatim notes). One thing to note is that a strong majority of students at Princeton reported that they generally took notes on a laptop, while a majority of UCLA students reported that they took notes longhand. It is heartening that our second study (run at UCLA) did replicate our first study (run at Princeton)."

265 the lecture's content In a note sent in response to fact-checking questions, Mueller wrote: "Laptop note-takers had far more content in their notes. Thus, we thought that the laptop note-takers' performance would rebound when they had a chance to look back on their notes—the laptop note takers just had so much more information available at the time of study. However (as we were quite surprised to find), it seems that if they didn't process the information at the time of encoding (i.e., during the lecture), the increased quantity of notes didn't help, or at least didn't help within a short study period. Perhaps with a longer time to study, they could piece together the content of the lecture, but at that point, the process is pretty inefficient, and it would be better to have taken 'better' (i.e., longhand-style, with less verbatim overlap) notes the first time around."

INDEX

Page numbers in *italics* refer to illustrations.

ABOUT THE TYPE

This book was set in Scala, a typeface designed by Martin Majoor in 1991. It was originally designed for a music company in the Netherlands and then was published by the international type house FSI FontShop. Its distinctive extended serifs add to the articulation of the letterforms to make it a very readable typeface.